HOW TO
REGISTER
YOUR OWN
TRADEMARK

with forms

HOW TO
REGISTER
YOUR OWN
TRADEMARK

with forms

3rd Edition

346.7304
W 256 h7

Mark Warda
Attorney at Law

SPHINX® PUBLISHING
A Division of Sourcebooks, Inc.®
Naperville, IL • Clearwater, FL

Third Edition, 2000

Published by: **Sphinx® Publishing, A Division of Sourcebooks, Inc.®**

Naperville Office
P.O. Box 4410
Naperville, Illinois 60567-4410
630-961-3900
FAX: 630-961-2168

Clearwater Office
P.O. Box 25
Clearwater, Florida 33757
727-587-0999
FAX: 727-586-5088

Cover Design: Sourcebooks, Inc.®
Interior Design: Sourcebooks, Inc.®
Interior Production: Mark Warda and Amy S. Hall, Sourcebooks, Inc.®

This publication is designed to provide accurate and authoritative information in regard to the subject matter covered. It is sold with the understanding that the publisher is not engaged in rendering legal, accounting, or other professional service. If legal advice or other expert assistance is required, the services of a competent professional person should be sought.

From a Declaration of Principles Jointly Adopted by a Committee of the American Bar Association and a Committee of Publishers and Associations

This product is not a substitute for legal advice.

Disclaimer required by Texas statutes

Library of Congress Cataloging-in-Publication Data
Warda, Mark
     How to register your own trademark : with forms / Mark Warda.—
3rd ed.
         p.   cm.
     Includes bibliographical references and index.
     ISBN 1-57248-104-8 (pbk.)
     1. Patents--Law and legislation--United States Popular works.
     2. Trademarks--Law and legislation--United States Forms.   I. Title.
KF3181.Z9W374   2000
346.7304'88--dc21
                                                                      99-41757
                                                                      CIP

Printed and bound in the United States of America.
HS Paperback — 10 9 8 7 6 5 4 3 2 1

CONTENTS

Using Self-Help Law Books

Before using a self-help law book, you should realize the advantages and disadvantages of doing your own legal work and the challenges and diligence that this requires.

A Growing Trend

Rest assured that you won't be the first or only person doing their own case. In some states, more than seventy-five percent of the people in divorces and other cases represent themselves. Because of the cost of legal services this is a major trend and many courts are struggling to make it easier. But some courts are not happy with people who do not use attorneys and refuse to help them in any way. For some the attitude is, "Go to the law library and figure it out for yourself."

We write and publish these books to give people an alternative to the obtuse law books in most law libraries. We have made the explanation as simple as possible so most people can understand it. But when writing a book, unlike when counseling a single client, we cannot cover every conceivable possibility.

Cost/Value Analysis

Whenever you shop for a product or service, you are faced with various levels of quality and price. In deciding what product or service to buy, you make a cost/value analysis on the basis of your willingness to pay and the quality you desire.

When buying a car, you decide whether you want transportation, comfort, status, or sex appeal. Accordingly, you decide among such choices as a Neon, a Lincoln, a Rolls Royce, or a Porsche. Before making a decision, you usually weigh the merits of each option against the cost.

When you get a headache, you can take a pain reliever (such as aspirin) or visit a medical specialist for a neurological examination. Given this choice, most people, of course, take a pain reliever, since it costs only pennies; whereas a medical examination costs hundreds of dollars and takes a lot of time. This is usually a logical choice because it is rare to need anything more than a pain reliever for a headache. But in some cases, a headache may indicate a brain tumor and failing to see a specialist right away can result in complications. Should everyone with a headache go to a specialist? Of course not, but people treating their own illnesses must realize that they are betting on the basis of their cost/value analysis of the situation. They are taking the most logical option.

The same cost/value analysis must be made when deciding to do one's own legal work. Many legal situations are very straightforward, requiring a simple form and no complicated analysis. Anyone with a little intelligence and a book of instructions can handle the matter without outside help.

But there is always the chance that complications are involved which only an attorney would notice. To simplify the law into a book like this, several legal cases must often be condensed into a single sentence or paragraph. Otherwise, the book would be several hundred pages long and too complicated for most people. However, this simplification necessarily leaves out many details and nuances that would apply to special or unusual situations. Also, there are many ways to interpret most legal questions. Your case may come before a judge who disagrees with the analysis of our authors.

Therefore, in deciding to use a self-help law book and to do your own legal work, you must realize that you are making a cost/value analysis. You have decided that the money you will save in doing it yourself outweighs the chance that your case will not turn out to your satisfaction. Most people handling their own simple legal matters never have a problem, but occasionally people find that it ended up costing them more to have an attorney straighten out the situation than it would have if they had hired an attorney in the beginning. Keep this in mind while handling your case, and be sure to consult an attorney if you feel you might need further guidance.

LOCAL RULES   The next thing to remember is that a book which covers the law in an entire state or the nation cannot possibly include every procedural difference of every jurisdiction. In some areas each county, or even each judge, is allowed to require their own forms or procedures. Our forms usually cover the majority of counties in a state, or are examples of the type of form which will be required. But keep in mind that your county or your judge may have a requirement not included in the book.

Before using the forms in a book like this you should check with your court clerk to see if there are any local rules or local forms which are required. Often these forms will require the same information as the forms in the book but are merely laid out differently. Sometimes they will require some additional information.

CHANGES IN   Besides being subject to local rules, the law is subject to change at any
THE LAW   time. The courts and the legislatures of all fifty states are constantly revising the laws. It is possible that while you are reading this book, some aspect of the law is being changed.

In most cases, the change will be of minimal significance. A form will be redesigned, more information will be required, or a waiting period will be extended. These should not affect the outcome of your case. You might need to redo a form, file an extra form, or wait out a longer time period. But in some cases, the entire law will be rewritten or a case you are relying on will be overruled.

Again, you should weigh the value of your case against the cost of an attorney and make a decision of what you feel is in your best interest.

INTRODUCTION

A trademark can be the most valuable asset a business owns. Imagine what would happen if just anyone could call their soft drink "Coca-Cola" or their hamburger restaurant "McDonald's." Those trademarks are worth millions of dollars in goodwill and repeat business to the companies that own them.

Many aspects of trademark law are considerably different in this edition of the book from the last two. The Internet has caused major changes in trademark law and will continue to do so for years to come. Many proposals for new laws are pending in Congress and courts are releasing new decisions each week spelling out the rights of trademark owners.

Also, the Trademark Law Treaty Implementation Act changed many of the procedures for registering a mark which took effect in late 1999. Instead of having strict rules and returning an application for any little discrepancy, the office now accepts nearly all applications and liberally allows corrections and clarifications.

To allow for scanning the forms into an optical character reader, the forms were also redesigned in 1999 and these new forms are included in this book.

The Federal Trademark Dilution Act now allows famous trademarks to keep even those in other fields from using similar marks. Decisions in recent court cases allow companies to stop others from using works that are similar in ways never before protected.

Trademark applications were up twenty percent in 1999 and are expected to soon pass 200,000 a year. Even a "capitalist tool" like *Forbes* magazine has claimed that the law is getting out of control with one company winning a Supreme Court case protecting its "tacky Mexican" restaurant decor and another obtaining a trademark for pink color for insulation.

While commentators may be right to decry these changes as limiting the rights of other businesses, smart entrepreneurs should take advantage of the changes to be sure not to lose any rights if the changes are upheld.

It is the purpose of this book to explain, in simple language, the steps necessary to protect your trademark by properly registering it. This book includes registration in the United States Patent and Trademark Office and in individual states. For years, the only books on the subject were legal treatises that were difficult even for lawyers to use. The first edition of this book was one of the first attempts to simplify this area of law for laymen.

It is advisable to read this entire book before attempting to register your mark. The first chapter explains the things you need to know about what a trademark is and what rights it grants. Chapter 2 explains the recent changes that the Internet has caused in trademark law. Chapters 3 through 6 explain how to prepare for filing your application. If you have not yet used your mark, you should use chapters 7 through 9 to register it. If you have already used it, use chapters 8 through 10.

In most cases, the registration of a simple trademark should go smoothly. If for any reason your application becomes complicated, you are urged to consult one of the treatises or an attorney who specializes in trademark law. Some good trademark books are listed in the bibliography to this text, but others may also be available at your local law library. Many of them do an excellent job of explaining each step in the complicated actions such as fighting an opposition to your application.

Occasionally, a company using a similar mark that learns of your application may threaten a lawsuit in Federal District Court for infringement of their mark. In such a case, unless you are willing to abandon the mark immediately, or dedicate much time to researching trademark law, it is advisable to work with a law firm specializing in trademark litigation to plan further strategy. This is explained in chapter 12.

FLOWCHART FOR REGISTERING A FEDERAL TRADEMARK

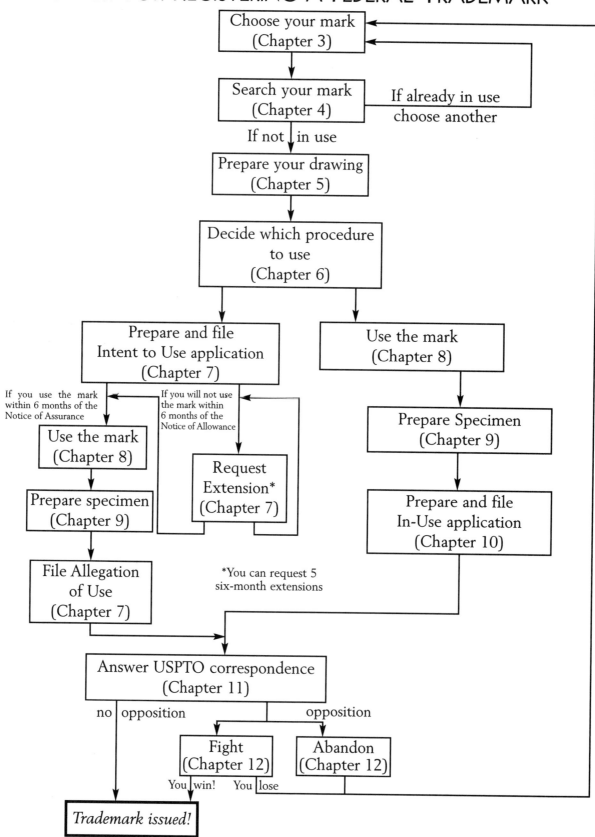

Trademark Basics 1

Differences between Trademarks, Copyrights, Patents and Trade Secrets

If you're baffled by the these terms and their differences, don't feel alone, even newspaper and magazine writers often confuse trademarks, copyright, and patents. But these, along with trade secrets, are four completely different types of protections, usually used for completely different types of property and protected in completely different ways. In order to obtain the right protection, you need to understand what each one protects.

PATENTS A *patent* is protection given to new and useful inventions, discoveries, and designs. A work must be completely new and "unobvious" to be entitled to a patent. A patent is granted to the first inventor who files for the patent. Once an invention is patented, no one else can make use of that invention, even if they discover it independently after a lifetime of research. A patent protects an invention for twenty years; and for designs, it is fourteen years. Patents cannot be renewed. The patent application must clearly explain how to make the invention so that when the patent expires, others will be able to freely make and use the invention. Patents are registered with the United States Patent and

Trademark Office (USPTO). Examples of patentable things would be mechanical devices or new drug formulas.

COPYRIGHTS

A *copyright* is protection given to "original works of authorship" such as written works, musical works, visual works, performance works, or computer software programs. You cannot copyright titles, names, slogans, or works that have not been fixed in tangible form. A copyright gives the author and his heirs exclusive right to his work for the life of the author plus seventy years. Copyrights are registered with the Register of Copyrights at the Library of Congress. Examples of works that would be copyrightable are books, paintings, sculptures, songs, poems, plays, drawings, and films.

TRADEMARKS

A *trademark* is protection given to a name or symbol that is used to distinguish one person's goods or services from those of others. It can consist of letters, numerals, packaging, labeling, musical notes, colors, or a combination of these. A trademark lasts indefinitely if it is used continuously and renewed properly. Trademarks are registered with the United States Patent and Trademark Office. Examples of trademarks are the "Chrysler" name on automobiles, the red border on TIME magazine, and the shape of the Coca-Cola bottle.

TRADE SECRETS

A *trade secret* is some information or process that provides a commercial advantage that is protected by keeping it a secret. Examples of trade secrets are a list of successful distributors, the formula for Coca-Cola, or a unique source code in a computer program. Trade secrets are not registered anywhere—they are protected by the fact that they are not disclosed. They are protected only for as long as they are kept secret. If you independently discover the formula for Coca-Cola tomorrow, you can freely market it. (But you can't use the trademark "Coca-Cola" on your product to market it.)

UNPROTECTABLE CREATIONS

Some things are just unprotectable. Such things as: ideas, systems, and discoveries are not allowed any protection under any law. If you have a great idea, such as selling packets of hangover medicine in bars, you can't stop others from doing the same thing. If you invent a new medicine, you can patent it; if you pick a distinctive name for it, you can register it as a trademark; if you create a unique picture or instructions for the package, you can copyright them; however, you cannot stop others from using your basic business idea of marketing hangover medicine in bars.

Notice the subtle differences between the protective systems available. If you invent something two days after someone else does, you cannot even use it yourself if the other person has patented it. But if you write the same poem as someone else, and neither of you copied the other, both of you can copyright the poem. If you patent something, you can have the exclusive rights to it for twenty years, but you must disclose how others can make it after the twenty years are through; however, if you keep it a trade secret, you have exclusive rights as long as no one learns the secret.

The following chart compares the important differences.

	Trademarks	Patents	Copyrights	Trade Secrets
Works Protected	Names or Symbols	Useful inventions, discoveries and designs	Written, visual musical or performance	Information or process
Requirements	Designation of origin	New and unobvious	Original work	Kept secret
Claimant	User	Inventor	Author	Owner
Duration	Unlimited if renewed	20 years	Life of the author plus 70 yrs.	Unlimited if kept secret
Protection if not registered	Yes	No	Yes	Yes
State Protection	Yes	No	No	Yes

TRADEMARK TERMS

Before attempting to use or register a trademark, it is important to understand the various terms that are used and the legal distinctions between them. The word "trademark" is used generally to describe several types of marks. These different types of marks are used for different purposes and, in some cases, different application forms are needed for registering them.

TRADEMARK   A *trademark* is a word, name, symbol or device (or any combination thereof) that someone uses on his or her *goods* to distinguish them from others' goods. "Trademark" is not the proper term when referring to a name, symbol, or device used in connection with services.

SERVICE MARK   A *service mark* is a trademark that is used in connection with *services* rather than goods. A service mark would apply to restaurant services, data services, or any other type of business that sells services rather than products. A restaurant (which is a service business) can also get trademarks on some of its products, such as the name Big Mac for a hamburger.

COLLECTIVE MARK   A *collective mark* is a word, name, symbol, or device (or any combination thereof) that is used by members of a group such as a union, trade association, or cooperative. For example, unions use the little "union bug" symbol to designate products made by their members.

CERTIFICATION MARK   A *certification mark* is a symbol indicating that goods or services meet certain criteria such as purity or approval. It is usually owned by one organization and licensed to manufacturers to use on their products. The Good Housekeeping Seal of Approval, the UL label, and the Real symbol on foods are examples of certification marks.

TRADE DRESS   A *trade dress* is an overall look of a product or company that is distinctive and identifies that product. Trade dress usually involves shape and color such as the Coca-Cola bottle or the red border on TIME magazine.

TRADE NAME   A *Trade Name*, the name used for a business, cannot be registered as a federal trademark because trademarks only apply to specific goods and services, not to company names. One way that a trade name can be protected is if it is registered as a trademark on goods, or a service mark on services, *and* used as a trade name. Another way trade names can be protected is through state law principles of unfair competition. See chapter 15 regarding other protection.

IN COMMERCE
In order to qualify for federal registration, a mark must have been used *in commerce*, which means in business dealings across state lines or with someone in a foreign country. For a small restaurant that wants to protect its name, this can require interstate advertising. This requirement is explained in more detail in chapter 7. If your business will be purely local and you will never want to expand to another state or to franchise the business, you do not need federal registration of your trademark. Chapter 14 explains how to register your mark in your state.

CLASSES
Goods and services are divided into classes, and marks are registered according to the class into which the goods and services fall. If you will use your mark on goods or services that fall into different classes, then you will have to register the mark in each class in order to protect it. You will have to pay an additional filing fee for each class (on January 10, 2000, it was raised to $325; to find out if the fee has changed, call 703-557-4636).

REQUIREMENTS FOR TRADEMARK PROTECTION

In order for a trademark to be protected by law, it must be used to *identify the source* of good or services. It is not enough that it is used as a name because, as explained above, trade names are not protected. This means the trademark must be displayed prominently on the goods or in advertisements for the services.

The goal is to get consumers to identify the name with the product or service. If ever a trademark is disputed and ends up in court, important evidence will be whether consumers identified the mark with the goods.

Two other important factors in determining whether a mark is a valid trademark is whether it is *inherently distinctive* or has acquired *secondary meaning*. Coined words like Exxon and Kodak are distinctive because they have never been used for any other purpose. Other words like General Electric might start out as generic, but become distinctive after years of advertising identify them with one particular company.

Some types of marks, such as slogans and surnames, cannot be registered unless it can be proven to the trademark office (through consumer surveys) that they have become inherently distinctive or have acquired secondary meaning.

TYPES OF TRADEMARK PROTECTION

There are three types of legal protections available for trademarks: common, federal, and state law.

COMMON LAW   The *common law* is the body of law contained in judges' decisions rendered over the centuries based upon societal principles of justice. Some of these principles protect businesses from those who take unfair commercial advantage of their name or brands. Because common law protection is based upon court decisions, it varies from state to state. You do not have to register anywhere for common law protection. If you are legitimately using a name, you are automatically granted certain rights just by using it.

The basic principle of most of these laws is that someone may not "palm off" his goods as those of someone else or misappropriate the efforts of others. What courts look for is a likelihood of customer confusion, which is against the public interest. It is not enough that two companies are competing for the same customers. There is nothing wrong with focusing on a profitable market segment and offering an alternative product. What you cannot do is fool the public into thinking your products are those of someone else.

Unfair competition law is an area of trademark law which is often used to protect trade names that are not entitled to trademark registration. That is, if a company adopts a name that is similar to another's name, and if the similarity causes customer confusion, then a court may forbid further use of the name. In some cases, a person may be prohibited from using his own name to market a product. For example, a person named Joe Rolex would not be able to sell watches using his last name.

FEDERAL TRADEMARK LAW
Federal trademark law provides a system for registering trademarks with the United States Patent and Trademark Office. It is contained in the Lanham Act and is the main focus of chapters 6 through 14 of this book. By using this system, you are given an easier way to prove your ownership and protect your rights. It offers the broadest protection for trademarks used in the United States.

STATE TRADEMARK LAW
State trademark laws have been passed by all of the states in the U.S. Unlike patents and copyrights, which are exclusively controlled by the federal government, trademarks may be registered with each state. The state laws offer simpler and less expensive registration, but the protection is limited to that state. If you plan to operate only a local business, you might want to consider state registration. More information is included in chapter 15.

TYPES OF MARKS

Different types of marks are entitled to be registered. These are:

WORDS
Words such as Kodak, Microsoft, Netscape or Omega.

SYMBOLS
Symbols such as the big red K used by Kellogg; the pentagon used by Chrysler; or the Jolly Green Giant figure.

NAMES
Names such as McDonald's, Chevrolet, or Templeton.

NUMERALS
Numerals or numeral combinations with letters or words, such as 747, Seagram's 7, or WD-40.

LETTERS	Letters, such as RCA for Radio Corporation of America, IBM for International Business Machines, or AA for American Airlines.
PACKAGING	Packaging, such as the shape of the Coca-Cola bottle or the Pierre Cardin cologne bottle. Companies have recently attempted to trademark the shape of a car as a trademark.
LABELING	Labeling, such as the red and white design on the Coca-Cola can.
SOUNDS	Sounds, such as the three musical notes used by NBC. Harley-Davidson is trying to trademark the sound of its motorcycle engines.
COLORS	Color, such as the pink color of certain fiberglass building insulation. Color is a difficult thing to trademark because there are a limited number of primary colors and if each company takes one, there will soon not be enough for new companies wanting to enter the field. Also, color cannot be trademarked if it is functional, that is, if the color is in some way important to use of the product. For example, a company that sold surgical instruments tried to trademark the color gray for its packaging. It lost because the court found that the color gray is useful as a background color to highlight the features of products in catalogs. *Specialty Surgical Instrumentation v. Phillips*, 30 U.S.P.Q.2d 1481 (8th Cir. 1994). However, in 1995 the U.S. Supreme Court ruled that a color can be trademarked if it has clearly become identified with a certain brand over time. *Qualitex Co. v. Jacobson Products Co., Inc.*, 115 S. Ct. 1300 (1995).
SLOGANS	Slogans can be used as trademarks, but they must usually be used for at least five years before they can be registered. Some examples of slogans are "You've come a long way baby" and "Just do it."
COMPOSITES	Composites of two or more of the above features, such as a Coke bottle in the registered shape with the registered logo in the registered colors.
LOOK AND FEEL	The "look and feel" of a product is a relatively recent attribute to obtain protection. Look and feel can include the interface of a computer program, the design of a sweater, or the ambiance of a restaurant. A recent case even held that the design of a golf course can be protected by a "look and feel" trademark and another course could not copy its layout.

NEW
TRADEMARKS

In addition to the above, companies are trying to trademark new things such as the moving image of a company logo at the beginning of a movie and the sound of a motorcycle engine. As the value of trademarks grows in the new information age, expect trademark law to also expand significantly.

TRADE DRESS

Trade dress is an expanding area of trademark rights. In recent years, lawyers have argued, and judges have accepted, that such things as clothing designs, computer interfaces, and even golf course layouts can be protected from infringement by trademark law. Today, attorneys are attempting to push the protection even further, claiming trademark protection for such things as a unique play used in a sporting event.

In a 1992 case, the U.S. Supreme Court took this to somewhat of an extreme. In that case, a restaurant chain sued another for copying its "tacky Mexican" decor. The decor was not even trademarked but the chain claimed that it was a distinctive look and that people identified their restaurant by that look. The Supreme Court bought it and said that an unregistered trade dress of a restaurant can be protected from others who would copy it. *Two Pesos v. Taco Cabana, Inc.*, 505 U.S. 763, 112 S.Ct. 2753 (1992).

This is not a good situation for a supposedly free market. Rather than lower its prices to fight competition, a large business can take a smaller one to court and force them to stop making a product that is too similar to theirs.

Large companies can fight back. In two recent cases, businesses tried to stop competitors from making competing flower pots and child carriers, and they convinced two federal courts they had a right to do so. Fortunately the competitors had the money to appeal, and the rulings were reversed. But how many small companies don't have money for legal battles and just give up? How many small businesses were forced to stop using names like "Insurance R Us" when intimidated by lawyers from Toys 'R' Us? (When a mark is new, it is only protected in the area of commerce in which it is registered. After it becomes famous, dilution law can protect it from similar names in any field.)

A more recent case indicates, however, that the important factor in deciding whether protection is given is not in protection of the trademark owner, but in protection of consumers from confusion. In this case, the maker of Vaseline Intensive Care Lotion came out with a new bottle and label for its product. A discounter copied the shape of the bottle as well as the colors and label, but also included its own logo and a notice that the product should be "compared to Vaseline Intensive Care Lotion." The court ruled that consumers would not be confused that it was the Vaseline product and that it was therefore not illegal. *Conopco v. May Dept. Stores*, 46 F 3rd 1556 (Fed. Cir. 1994). As one commentator put it, the decision struck terror into the hearts of trademark lawyers.

An important issue in trade dress cases is whether the look of the product was used to identify the source of the goods. In many cases, the designs, on clothing for example, merely create a style and not an indication of the source of the product. A recent case, *Samara Brothers, Inc. v. Wal-Mart Stores Inc.*, 165 F3d 120 (2nd Cir. Dec. 28, 1998), allowed a clothing maker to collect damages from Wal-Mart for selling similar designs. This was a break from previous cases. However, one of the three judges strongly dissented and it is possible that this case will not be followed.

Trade dress protection has been used to keep competing websites from looking too similar. This means that you should not design your site to have the same "look and feel" of a competitor's site and if your competitor copies your site you may have grounds to challenge him.

If you have a unique creation of any type that you are not sure is protectable, consider fitting it into one of the trademark categories. However, understand that this is one of the most confused areas of trademark law. Consider that after the Supreme Court's decision on Mexican restaurants, some circuit courts ignored it and made up their own test, other circuits criticized them, and district court judges criticized circuit court opinions. So if federal judges can't make up their minds on what the law is, there is little chance even the best lawyer will know.

The bottom line is, try to get as wide a trademark as you can, but be prepared for an expensive legal battle if a larger company goes after you.

TYPES OF FEDERAL REGISTRATION

The Patent and Trademark Office maintains two different registers of trademarks, the Principal Register and the Supplemental Register.

THE PRINCIPAL REGISTER

The *Principal Register* is the one that provides all of the legal rights explained under REASONS FOR REGISTERING A TRADEMARK that follows.

THE SUPPLEMENTAL REGISTER

The *Supplemental Register* is used for marks that cannot be registered in the Principal Register because they are descriptive, geographic, or surnames. Registration on the Supplemental Register prevents federal registration of the mark by others and can later register on the Principal Register after the mark becomes distinctive. It does not offer most of the other rights which the Principal Register does.

If you have applied for registration on the Principal Register and the examining attorney rules that your mark cannot be registered, for example, because it is a surname, you may be allowed to amend your application to seek registration on the Supplemental Register.

FORBIDDEN TRADEMARKS

Certain marks may not legally be registered, and it is a waste of time and money to try. Some that are forbidden are:

GENERIC

Generic names are names that merely describe a product. For example, Blue-Denim Jeans could not be registered since it describes a product. But you could call your product "Klondike Blue Denim Jeans" and register the name "Klondike" (unless someone already has) (15 U.S.C. § 1127).

Foreign words which are generic are not allowed to be registered. In a recent case, the trademark Otokoyama for saki was thrown out because a competitor proved that otokoyama was a type of saki sold by several companies in Japan. So one company was not allowed to keep others from describing their saki as being that type.

IMMORAL OR SCANDALOUS

Immoral or *scandalous* has been defined to include things that are shocking to the sense of propriety, offensive to the conscience or moral feelings, calling out for condemnation, vulgar, lacking in taste, indelicate, or morally crude. For example, a photograph of a nude couple embracing and kissing was rejected, as was the mark "Bullshit." However, "Weekend Sex" was accepted for the name of a magazine, and "Big Pecker Brand" was acceptable as a trademark for T-shirts since it included the picture of a bird which "would make it less likely that purchasers would attribute any vulgar connotation to the mark" [15 U.S.C. § 1052(a)].

FALSE CONNECTION

A mark may not falsely suggest a connection to a person or institution. Besides being unregisterable, using someone's name on a product would violate their right to privacy for which they could sue for damages. Even using a person's nickname or other designation could be a problem. Chi-Chi's restaurants were sued by Jimmy Buffett for using the term "Margaritaville" for which he was known [15 U.S.C. § 1052(a)].

DECEPTIVE

A mark may not be deceptive. For example, "Lovee Lamb" was deceptive when used on seat covers not made of lambskin, "Perry New York" was deceptive when used on clothing originating in North Carolina, and "American Limoges" was deceptive when used on dinnerware neither from Limoges, France, nor made from Limoges clay. However, the mark "Sweden" was acceptable for artificial kidney machines though they were not made in Sweden because the deception was held to be "perfectly innocent, harmless or innocent" [15 U.S.C. § 1052(a)].

DISPARAGING

Marks that disparage a person, institution, belief, or national symbol are not registerable. It is not a problem to use a national symbol, only to use it disparagingly. National symbols have been held to include such things as the bald eagle and hammer and sickle (however since the collapse of the Soviet Union, the hammer and sickle may now be available, though perhaps not valuable, as a trademark). Things such as the "House of Windsor," the space shuttle, or the Boston tea party have been held not to be national symbols [15 U.S.C. § 1052(a)].

In 1999, the USPTO refused to register the mark "Slick Willie" for cigars because it would be disparaging to the president of the United States.

FLAGS, INSIGNIA, OR COATS OF ARMS
A private party may not trademark the flag, insignia, or coat of arms of the United States or any state, municipality, or foreign nation. Also forbidden is a simulation of any of these. Whether a mark simulates a flag or coat of arms depends upon the exact characteristics of each. An eagle in a triangular shield, not held to simulate the seal of the United States, and a globe with six indistinguishable flags were both acceptable. The letters USMC were held not to be an insignia of the U.S. Marine Corps. and were allowed as a trademark, but a special law was passed to be sure that this wouldn't happen again [15 U.S.C. § 1052(b)].

NAME, PORTRAIT, OR SIGNATURE
One may not trademark the name, portrait, or signature of a living person without his or her written consent or of a deceased president of the United States during the lifetime of his widow without her written consent. For example, Steak and Ale Restaurants were not allowed to register "Prince Charles" as a brand of steak (even though there may be many persons with such a name), but Coca Cola was allowed to register the drink name "Fanta" over the objections of Robert D. Fanta [15 U.S.C. § 1052(c)]. (See chapter 3 for more comparisons.)

CONFUSINGLY SIMILAR
A mark may not be registered if it is confusingly similar to a registered mark or a mark previously used by another that has not been abandoned. Through your search, you may notice marks similar to your own. Even if they are not exactly the same as your mark, yours may be rejected if the examiner thinks that the public would be confused. Examples of marks that were too similar are, "Trucool" and "Turcool," "Commcash" and "Communicash," "Cresco" and "Kressco," "Entelec" and "Intelect," "Seycos" and "Seiko." Examples of marks that were held to be not confusing are, "Tia Maria" and "Aunt Mary's," "Best Jewelry" and

"Jeweler's Best," and "Cobbler's Outlet" and "California Cobbler" [15 U.S.C. § 1052(d)].

GEOGRAPHIC   A mark may not be registered if it is primarily geographic. For example, you cannot register as a trademark the name "Georgia Peaches." But you could call your peaches "Sunshine Georgia Peaches" and register the word "Sunshine" (unless it has already been registered) [15 U.S.C. § 1052(e)].

SURNAMES   Unless a surname has acquired distinctiveness it may not be registered. This is because one person cannot establish exclusive rights to his surname until the public has begun to identify the name with the goods or services. One well-known surname which has acquired distinctiveness is McDonald's [15 U.S.C. § 1052(e)(3)].

FUNCTIONAL FEATURES   If the feature of a product is functional, that feature may not be trademarked. This is so that competitors are not stopped from making the product. For example, Rolodex Corp. tried to trademark index cards with rounded corners and curved sides; and Bose Corp. tried to trademark a five-sided loudspeaker enclosure. Both were refused (15 U.S.C. §§ 1051, 1052 and 1127).

ORNAMENTATION   Mere ornamentation that does not *distinguish* the goods from other goods may not be registered, since the point of a trademark is to distinguish goods or services (15 U.S.C. §§ 1051, 1052 and 1127).

OLYMPIC   The word OLYMPIC and the olympic logo were the subject of a special law passed by Congress. No one may use these even on an unrelated product or service. This law was upheld as constitutional by the U.S. Supreme Court. *San Francisco Arts & Athletics, Inc. v. United States Olympic Committee*, 483 U.S. 522 (1987).

TRADEMARK DILUTION

As explained earlier, a trademark is registered by class and only covers goods in one class. For example, in the past, the Ford Motor Company could not stop someone from using the trademark Ford for something unrelated—like bath oil beads. However, as corporations have expanded their operations into other fields and into other countries, they have sought to limit others' use of marks which are similar to their own.

A way companies have protected their marks from competitors in other fields is to claim that they *diluted* the value of their mark. Some states have laws which forbid uses which would dilute the value of a mark or tarnish its image. This theory only applies to famous marks that much of the population recognizes. For example, even though Coca-Cola is a registered trademark for beverages, if someone tried to sell "Coca-Cola" brand shoes they would probably be ordered by a court to stop because they would be diluting the value of the mark to the Coca-Cola Company (and would confuse the public).

An important court case on this area of law is *Mead Data Central Inc. v. Toyota Motors Sales U.S.A. Inc.*, 875 F. 2d 1026 (2d Cir. 1989). In that case, Mead sued Toyota to try to stop them from calling their new car Lexus because they said it would dilute the value of Mead's trademark Lexis for their legal information service. The district court ruled for Mead, but on appeal, the decision was reversed. In a long and thoughtful decision, the court explained the law regarding dilution. The six factors that the court said were important in a dilution-by-blurring case are:

☛ similarity of the marks

☛ similarity of the products

☛ sophistication of the customers

☛ predatory intent of the second mark's owner

☛ renown of the first mark

☛ renown of the second mark

FEDERAL
TRADEMARK
DILUTION ACT

In 1995, Congress passed a law intended to give companies that own famous trademarks more protection from those who would use similar marks. This law allows companies to stop others from using marks that cause dilution of their marks regardless of whether the companies are in competition or whether there is consumer confusion. This is much greater protection than the law previously allowed.

However, despite the intent of the statute, most courts are still using the *Mead* test and requiring some basis for customer confusion.

Because the statute's protections may offer protections so broad that they violate the rights of others, the courts' limitations may be good in that they keep that act from being overbroad or even unconstitutional.

Here are a few examples of the results of the early cases under the law.

☛ Ringling Brothers sued a bar that billed itself as the "Greatest Bar on Earth" and the Utah Division of Travel Development for claiming it had the "Greatest Snow on Earth" claiming they diluted their mark "Greatest Show on Earth." Ringling lost both cases. However, years earlier, Ringling was able to stop a car dealer from advertising the "Greatest Used Car Show on Earth."

☛ American Express sued a company using the slogan "Don't Leave Home Without My Pocket Address Book," claiming it infringed its mark "Don't Leave Home Without It" and lost.

☛ Anheuser-Busch was able to stop a company from selling t-shirts with the logo "Buttwiser" on them.

☛ Toys 'R' Us was able to stop a company from calling an adult-oriented website "Adultsrus."

You may notice a pattern here. It appears that courts are less likely to allow a similar mark if the use appeals to prurient interests.

The important thing to remember when choosing your mark is that it should not be so similar to a famous mark that the owner of that mark will sue you. In a close case, where your products are different you may win, but can you afford a lengthy court fight with a major corporation? Of course, if you have a lawyer in your family who would help you at a reduced rate, you might get a lot of publicity for your company by fighting a Goliath.

REASONS FOR REGISTERING A TRADEMARK

A trademark is legally entitled to some protection even if it is not registered. As explained previously, the common law recognizes that a person who uses a distinctive mark on his goods is entitled to court protection from those who try to pass off their goods as theirs. However, the registration of a mark under federal law entitles the registrant to the following additional legal rights and benefits:

PRIORITY   The benefit of nationwide priority. This means that even if you are using the mark in only one area, registration gives you the right to the mark in every area of the country where it is not already being used.

CONSTRUCTIVE NOTICE   The benefit of *constructive notice*. This means that no one can say they didn't know you were using the mark. Registration officially notifies everyone in this country that you own the mark, whether they actually know it or not.

FEDERAL COURT   The right to sue in Federal District Courts for trademark infringement.

DAMAGES   The right to recover profits, triple damages, court costs, and attorneys' fees in a court action.

CUSTOMS   The right to deposit the registration with the Customs service to stop the importation of goods that bear an infringing mark.

EVIDENCE OF VALIDITY   The benefit of *prima facie* evidence of validity of your mark. This means that if you go to court, you don't have to waste time and money proving your ownership and exclusive right to use the mark.

INCONTESTABILITY   The benefit of possible incontestability, which means that the registration is conclusive evidence of your exclusive right to use the mark in commerce.

STRENGTH OF MARK   The benefit of more limited grounds for attacking the registration once it is five years old.

CRIMINAL PENALTIES   Another benefit is that there are criminal penalties for those who may counterfeit your mark. The penalty may be up to $250,000 or five years for the first offense, $1,000,000 or fifteen years in prison for the second or subsequent offense.

FOREIGN REGISTRATIONS

You have the benefit of having a basis for filing trademark applications in foreign countries.

In addition to the legal rights, the registration of a trademark confers some practical benefits:

ASSET

Registration makes the trademark a quantifiable asset that can more easily be licensed, sold, mortgaged, or transferred.

TAX BENEFITS

Registration allows some tax benefits both for Federal income tax purposes and for some state tax purposes such as service taxes.

SALABLE

If a large company someday wishes to use your mark, they may offer to pay you for your rights in it.

ACTIVITIES THAT ARE NOT PROTECTED

Although the owner of a trademark is guaranteed "exclusive use" of the mark, that right is limited as follows:

PREVIOUS USERS

Registering a mark does not allow you to stop others who have previously been using the mark. Persons who have used the mark prior to the person who registered it may continue to use it in the area where they have used it in the past. However, they may not expand their use to other areas of the United States once it has been registered by someone else.

OTHER CLASS

A registration only covers a specific class of goods. For example, if one company registered the mark Penguin for tuxedos, another company could register the same mark for ice cream bars. However, once a mark becomes famous dilution law protect it from similar marks in any classes.

OWNERS OF THE PRODUCT

An owner of a product can use the trademark of the goods he is selling or leasing. If you own a Mercedes-Benz, you can use that trademark to advertise and sell the car you own.

COMPETITORS' COMPARISONS
A competitor may use another person's registered mark in a comparison of the goods. For example, Carlton cigarettes can use the Camel cigarette trademark in an ad that states that Carlton has lower tar.

SCENES
Someone taking a photo or making a film of a scene comprising a variety of objects can include trademarked products without permission. For example, someone taking a photo of Times Square does not have to get permission for all of the trademarks visible in the scene. The reason for this is common sense—outdoor photographers would be out of business if they had to get permission for every mark in their photos. In reality, many trademark owners eagerly pay filmmakers to use their marks and this has become a big source of revenue for them. When you see Sylvester Stallone using a certain brand of cigarettes, it is not because it is his normal brand, it is because a company paid well for the placement of its product in his film.

PARODIES
While trademark law does not contain the same specific permission for parody that copyright law does, First Amendment principles and the general confusion between copyright and trademark law have resulted in rulings that allow trademarks to be used in parodies. As the Seventh Circuit Court of Appeals noted, "When businesses seek the national spotlight, part of the territory includes accepting a certain amount of ridicule." *Nike, Inc. v. "Just Dit It" Enters.*, 6 F. 3d 1225 at 1227 (7th Cir. 1993).

FOR FURTHER GUIDANCE

THE INTERNET
The United States Patent and Trademark Office has a website that includes a lot of useful material, including the latest rulings and revisions of forms. The address is http://www.uspto.gov.

TREATISES
For much more detailed explanation of the fine points of trademark law, a treatise on the subject would answer nearly any question. These are available at larger law libraries. The best law libraries are located at law schools and most are open to the public. See the books listed in the Bibliography.

T.M.E.P.   The *Trademark Manual of Examining Procedure* (TMEP) is the book used by government trademark attorneys to review your application. It answers nearly any question that could arise in the process. If your application is complicated or you want to know more about how your application will be examined, you should review it.

It is available online at http://www.uspto.gov/web/offices/tac/tmep/ or can be ordered from the Superintendent of Documents. The edition current at time of publication was $44 but check if it has increased. The order number is 903-010-00000-2. You can order it with a credit card by phone at 202-512-1800, by fax to 202-512-2250, or by mail from the following address:

> Superintendent of Documents
> P. O. Box 371954
> Pittsburgh, PA 15250-7954

Note: At the time of publication the T.M.E.P. was not up to date because of major changes to the law which took effect in October 1999. For the latest information, see: http://www.uspto.gov/web/offices/com/sol/notices/fr990908.htm.

The following websites contain some useful information on trademark law:

http://www.ggmark.com

http://www.phillipsnizer.com

http://www.fplc.edu/tfield/Trademk.htm

TRADEMARKS IN THE INTERNET AGE 2

The advent of the Internet has caused many changes in trademark law. From the type of search which must be done to the method of choosing a mark, the procedures are much different today than they were just a few years ago. This chapter summarizes the major changes. Most of these issues are explained in much more detail in other chapters.

DOMAIN NAMES

Domain names are the "addresses" of company sites on the Internet. For example, to find the site of the New York Times, you would go to www.nytimes.com. When the Internet first started, companies that wanted to use or reserve a specific name needed only to pay $100 and register it with Network Solutions, Inc., a company that was licensed by the U.S. Government's National Science Foundation.

The law of domain names has not yet been completely developed, but there are many laws and court cases pending at this moment that will begin to shape those rights. Rest assured that there is a lot of money being used to protect all of the vested interests.

Early in the development of the Internet, certain people in the computer industry who saw the importance of domain name rights to major corporations registered as many names as they could think of that might be important to some corporation. Some invested as much as hundreds of thousands of dollars to register thousands of names.

This worked in some cases, and when companies finally decided to set up a website and found that the name they wanted most was taken, they paid the original registrant for the rights.

But at least one company decided to fight the registrant. Intermatic, Inc. is the owner of a trademark "intermatic" for electrical timing devices. When it attempted to register "intermatic.com" as its domain name, it found that a Mr. Toeppen had already done so. In fact, Mr. Toeppen had registered about 240 domain names including such names as "nieman-marcus.com," "britishairways.com" and "ussteel.com." He offered to sell the name to Intermatic. But rather than purchase the name from Mr. Toeppen, Intermatic sued, claiming that his use was in infringement of their trademark. While there was no statute covering domain names, no clear precedent in the case law and no proof that Mr. Toeppen had actually used the trademark in a commercial way, the court analyzed the situation in a thirty-two page opinion. The court decided that use of a domain name is commercial in and of itself, and ordered Mr. Toeppen to hand over the domain name to Intermatic. *Intermatic, Inc. v. Toeppen*, 40 U.S.P.Q.2d 1412, 96 Civ. 1982 (N.D. Ill. 1996).

Some people even started registering the names of famous people such as sports and entertainment figures and someone offered Cal Ripkin the domain calripkin.com for $100,000. To avoid paying this extortion the more clever people registered a variation of their name such as by adding their jersey number or "inc" to their name, such as monicalewinskyinc.com.

In 1999, Congress passed the Trademark Cyberpiracy Prevention Act. This law allows companies which own trademarks, and people with famous names, to take back their marks and names and obtain damages from people who registered them in bad faith. Bad faith would be someone who registered a name without a clear intent to use it for a business, but who apparently intended to sell it or hold it for ransom. Since intent is in the mind of the participant, this would be a legal question for a judge or jury to decide.

In the meantime, the contract with NSI has not been renewed and over eighty organizations are cooperating to develop a new domain name policy. At one point they agreed to create seven new generic top level domains (gTLDs) and to set up at least twenty-eight new registrars around the world to accept registrations. The new gTLDs would be .firm, .store, .web, .arts, .rec, .nom, and .info. However, only two new gTLD had been made available by 1999—.cc and .tv. This is just the beginning, as even more gTLD names are expected in the future, and eventually .com will just be one of many possible addresses.

This will most likely free up a great number of names. However, policies proposed by the World Intellectual Property Organization (WIPO) would allow owners of famous trademarks to prevent anyone from using their mark in another domain. One problem with this law is that even if one famous company has a name trademarked, others may have legal rights to it. For example, there are several registered trademarks using the marks "FORD" and "MCDONALD" other than the ones which first come to mind. This rule might prevent any of those other companies from using their trademark in another gTLD if Ford Motor Company or McDonald's Restaurants objected. It is likely that it will be many years before the rights of the various parties are ultimately defined.

MISSPELLINGS   Another issue in the domain name arena is the misspelling of names that are typed in by searchers. For example, a competitor of the pest control company Terminix might register the domain name "terminex.com" so that anyone misspelling the name would go to their site. They would then offer competing services and attempt to win the customer. Some companies list several misspellings of their competitors' names in hidden meta tags to lure erroneous hits.

One way to avoid this problem when setting up a website is to register all the possible misspellings of your name so that they are not available to others. Some political candidates have bought names which could become embarrassing to them, such as "bushsucks.com" or "nogore.com" before anyone else could. One problem is that there are so many variations that you could never buy them all. For example, you could add "2000" to either of the above names, or register them under ".net" or ".org" or ".cc" and so on.

If you have enough funds, you might want to purchase the most common misspellings of your domain name and then have those sites redirect people to your site. One company found that twenty-five percent of their website hits came from people who misspelled their name and were redirected!

TRADEMARKING
A DOMAIN

A common question about domain names is whether they can be trademarked solely for use as a domain name. The answer of the Trademark Office is that if a domain name is used only as a website address, it cannot be registered. The analogy is to a phone number. It is merely a way of reaching you, it is not a symbol identifying your goods or services.

However, if the domain name is used to identify the source of the goods or services as well as being a domain name, it can be registered. This means it must be used on labels on the goods or in advertising for the services. If your services are provided solely through your website, then the name should be prominently displayed on your opening screen.

The Trademark Office has taken the position that adding ".com" to a descriptive or generic word (which would not be registrable by itself) does not make it registrable. Thus "insurance.com" could not be registered as a trademark.

CHOOSING YOUR MARK

Because the Internet reaches every corner of the world, you must now consider that every other company on the Internet may be your competitor. And there is now more of a likelihood that a common name you choose would infringe someone else's use of it somewhere. Therefore, it is more important than ever to either coin your own word which never before existed (like Exxon or Kodak) or to use some unusual combination of words which no one else would have thought of using (like Waikiki Snowshoes or Cellar in the Sky). The next chapter explains how to choose your mark.

Searching Your Mark

Prior to the Internet, a business could decide if it wanted to build a national trademark or to stay local. If it wished to stay local, it could register only at the state level or not register at all. The search could be of only local businesses.

Today, every business with a website has a presence in every corner of the globe and every other website on earth is available in your town. This means that even if you run a childcare business in your small neighborhood, your name could infringe a national chain of child care centers that doesn't yet operate in your state if they have a website. Or your one-person T-shirt printing operation might infringe a printer in India.

The full legal implications of the Internet are not yet known. Congress is considering numerous proposals, but there is not yet a consensus. Courts are deciding peoples' rights, but the decisions sometimes conflict. The best we can do at this point is give you some guidance on what would be the safest tactics in this unsure situation.

Before adopting a name as your trademark, besides a normal trademark search, you should search it in as many search engines on the Internet as possible. If you find another business using it, you should consider seeking another name. If you find it used on other website, but only noncommercial pages such as personal websites, you would probably be safe in using it as a trademark. Chapter 4 explains more about searching your mark.

REGISTERING YOUR MARK

You can now submit your trademark application twenty-four hours a day, seven days a week, over the Internet. You can pay with a credit card (AmEx, Discover, MasterCard, or Visa) and submit the required specimens in the form of JPG or GIF images.

The application, however, only works with Netscape Navigator 3.0 or higher, or Microsoft Internet Explorer 4.0 or higher. It will not work on the Macintosh with Internet Explorer.

The site for online registration is http://uspto.gov/teas/index.html. If you do not wish to file online, the site also allows you to prepare and print your application for mailing.

USING TRADEMARKS ON YOUR INTERNET SITE

If you have a website, you should be aware of the laws governing the use of others' trademarks on your site. The rules are generally the same as if you were publishing a newspaper. You can use trademarks to truthfully describe or sell products and for parody.

But the nature of websites allows other uses. For example some sites have taken other companies' names and trademarks and hidden them on their site to attract searchers. (A hidden search word is called a meta tag.)

For example, a maker of a cola-flavored soft drink might hide the words "Pepsi" and "Coke" in their meta tags so that anyone searching for those products would find this competitor's site. So far most courts have ruled that this is infringement and many companies have hired people to scour the Internet for infringing uses of their trademarks.

TRADE DRESS OF YOUR INTERNET SITE

As explained in the previous chapter, the "look and feel" of a product can be protected by trademark law. As it applies to your Internet site, the law protects you from competitors who make their sites too similar to yours and vice versa.

While the nature of the computer screen limits the possible layouts of a page, and no one can trademark a simple function of a page, unique design and layout can be protected. This is especially true if the person copying is in the same field of commerce.

NEW FILING PROCEDURES

It is now possible to file a trademark in a matter of minutes online. The fee can be paid with a credit card and the samples can be sent in later. Also, the paper form for registering a trademark has been revised so that it may be scanned into the Trademark Office computers and converted to text more easily.

CHOOSING YOUR MARK 3

Before spending the money to search a mark and file your application for registration, you should take the time to carefully choose a mark that will be both practical and legal. You should have already read the previous sections on what types of marks are available (page 15) and what types of marks are forbidden, (pages 19-20).

The two most important rules in choosing a mark are to be sure it is *not descriptive* and will *not cause confusion* with other marks. Other factors to consider are whether it is easy to remember, what images it evokes and, if you are exporting, what it means in foreign languages. (General Motors learned the importance of checking its marks in other languages when it realized that the name of its Nova model meant "It doesn't go" in Spanish!)

TYPES OF MARKS

There are four basic types of words that can be used to identify goods and services: coined, arbitrary, suggestive, and descriptive.

COINED   The best types of marks are those which are created and have no prior meaning. Kodak is a good example. The word had no meaning until Mr. Eastman started using it for his products. Therefore, it could not fall into any of the forbidden categories and no one else would have been using it. Another more recent example is Exxon, which was the new name adopted by the Esso company. Because Esso was doing business in so many countries, it wanted a mark that would be useful all over the world and not have any negative meaning. It used a computer program to create the word Exxon and to be sure it had no prior meaning in any language in the world.

However, one drawback to using a newly coined word for a mark is the expense of teaching consumers what it means. Jaguar or LeBaron gives you an impression of what type of car you are supposed to be buying. But if an automobile company started marketing a Qqueezoo car, what image would it inspire without a massive advertising campaign?

ARBITRARY   The second best type of mark is one that is arbitrary and unrelated to the product, such as Domino's pizza or Peaches record stores. With such a mark, it is less likely that someone else is using the same mark on a similar product, and it is in no way descriptive of the product.

SUGGESTIVE   A less advantageous type of mark is one that suggests the nature of the product. Examples of this would be Mr. Jiffy printing or Coppertone suntan lotion. People like to use these kinds of marks because they are easy to remember and often suggest that the goods have an advantageous quality. The disadvantage of using these over arbitrary and coined marks is that there is likely someone out there somewhere who is already using the mark.

DESCRIPTIVE   The weakest mark would be one that is descriptive of the product. Examples would be Long Life batteries or Krispy crackers. If a mark is "merely descriptive" it won't even be registered unless it can be proven that after extensive use, the mark has acquired a secondary meaning. This means that a large number of persons identify with the applicant's product when they see the mark. Unless you have been using a descriptive mark for many years, you should not try to register it as a trademark.

Choosing a Mark

Because a trademark is something that will identify your product and probably be used for the life of your business, you should choose it carefully. You should start with a list of several possible marks because, with the millions of businesses in this country, it is likely that some of the marks you think are best are already taken.

If you are about to launch a blockbuster product or service that will take the nation by storm, or if you have a large marketing budget, then you can be less careful in choosing your mark because the product or advertising will give recognition to it. But if you would like your mark to help your product or service, it should be especially clever or distinctive so that people remember it.

An example of a cute name would be "curl up and dye" for a beauty parlor. A name like that would likely get mentioned in a local paper and perhaps even in a national publication. A slightly more obscure name such as "edifice wrecks" for a demolition company would attract the attention of many people, but a large segment of the population just wouldn't get it.

Using a mark like Domino for a pizza restaurant or, even better, Subway for a submarine sandwich restaurant, opens up the possibility of using distinctive domino or subway graphics in the restaurant, its advertising, and promotional materials.

So, how do you choose a mark? Sit down with a dictionary, an encyclopedia, and a thesaurus and start looking up words you like or that might work with your product. The creators of early page layout programs for the Macintosh named their company Aldus after an early pioneer in the printing field.

One way to expand possible trademarks is to use more than one word for your mark. For example, while you can't use Georgia or "peaches" alone as a trademark, you might want to use the mark "Sunshine

Georgia Peaches" as a trademark. However, you will only be able to claim the word "sunshine" (since "Georgia" and "Peaches" cannot be registered, as explained in the first chapter). What you can do is use "Sunshine Georgia Peaches" as your mark and to disclaim any rights in the other two words. This is done by including in the application a statement such as, "No claim is made to the exclusive right to use 'Georgia' or 'peaches' apart from the mark as shown."

SOFTWARE   Computer software is available that is supposed to help the process of selecting a name. While it is pricey for a one-time use, you may wish to consider it if you need to name several products.

CONSULTANTS   There are consultants who offer to help find an ideal trademark for a fee; however, the fee is often a lot more than a small business can afford. A business magazine once ran a story about a major corporation that paid $50,000 for a company to create a name for their new breakfast cereal. The magazine also polled its own staff for suggestions based on the criteria given to the consultant and came up with the same name!

As explained in the last chapter, the advent of the Internet now requires even more care in choosing a mark and it is even more important that the mark be either coined or fanciful.

USING A STYLIZED TRADEMARK

Besides the basic word that usually constitutes a trademark, a stylized mark or logo may be registered. The advantage is that you can keep others from using similar designs, the disadvantage is that they may be able to use a similar word in a different style and not be found to be confusing. Also, the search is more difficult and expensive.

CONFLICTING MARKS

With all the millions of businesses in this country, and the countless products and services produced by each, it is unlikely that your mark will be unique and not in use by anyone anywhere. The question is how similar your mark can be to another mark.

Unless you think you can benefit from a David and Goliath lawsuit with a major corporation (and can afford it), you should not use a mark that is similar to a nationally known brand. The new federal anti-dilution statute gives owners of famous brands the power to stop others whose marks come close to theirs. You may not get noticed by a major company for several years, but if they do pursue you after you have been using a mark for several years, it may be very expensive to make a change. So avoid a mark that looks like a national brand.

As this book goes to press the *Wall Street Journal* has made a demand that the *Small Street Journal*, a Maine newspaper with a circulation of 7,000 which is given away free to children, change its name. At least one of the Wall Street Journal's lawyers thinks the tabloid is a threat to their legal rights, but their good name may suffer more from the bad publicity of this case.

EXACT
MATCHES

If you find someone using the exact mark you wish to use, you should first get answers to the following questions:

- ☛ Is it being used as a trademark?

- ☛ Is it being used for a business purpose?

- ☛ Is it being used in the same field of commerce?

- ☛ Is it a national brand?

- ☛ Might there be a conflict in the future?

If the mark is not used for a business purpose, you have no worry. Suppose you feel that Saharan Navy is a clever name for a trademark for boat equipment. You search the USPTO Database and find that no one is using it. Then you use a few Internet search engines and find a site which mentions a Saharan Navy in science fiction stories based on a future African flood. The site is run by a university student who writes the stories as a hobby and posts them for the world to read. Well, since the term is not being used as a trademark, and even if it were the name of the website it is not used for a commercial purpose, you are safe to register it as a mark.

Even if in the future the student wanted to trademark the name for his science fiction stories, it would not conflict with your use because boating equipment is in a completely different field of commerce. However, if you searched the mark Coca-Cola and found that it was not registered in the field of boating equipment, you would not be able to register or even use it. Because the mark is so famous, any use of it in any field would be infringement. See Trademark Dilution in chapter 1.

If your search found that someone had been using Saharan Navy for scuba equipment, you would probably be able to get a trademark for it if your boating equipment was unrelated to scuba diving. But the fact that it is being used in a field which is so close makes it possible that as either of your businesses expands, there will some day be a conflict. You should choose a different mark if you wish to avoid potential conflict. If your business will be much bigger and the name is important to your plan, you can go with it, realizing the potential for future problems.

SIMILAR
MARKS

If you do a trademark search and find no exact matches, but some that are similar, either in sound or in spelling, you must analyze the likelihood of confusion and weigh the chance that your mark will be rejected either by the trademark office or by a court if you are sued by the other owner.

This is the type of legal analysis that an experienced trademark lawyer can do much better than a layman.

Examples of similar marks that have been rejected.

One of the following was rejected because the other already existed.

- CONFIRM for blood gas analyzer and CONFIRMCELLS for diagnostic blood reagents.

- LAREDO for land vehicles and LAREDO for pneumatic tires.

- BIGG'S for a grocery and general merchandise store and BIGGS for furniture.

- GOLDEN GRIDDLE for pancake house restaurant and GOLDEN GRIDDLE for table syrup.

- MUCKY DUCK for mustard and THE MUCKY DUCK for restaurant.

- CAREER IMAGE for women's clothing and CREST CAREER IMAGES for uniforms.

- 21 CLUB for clothing and THE "21" CLUB for restaurant services and towels.

- VEGETABLE SVELTES for wheat crackers and SVELTE for frozen dessert.

- SPRAYZON for industrial cleansers and SPRA-ON for furniture cleaner.

- RESPONSE for banking services and RESPONSE CARD for twenty-four hour bank card.

- SEYCOS for watches and SEIKO for watches.

- CAYNA for soft drinks and CANA for fruit and vegetable juices.

- BUENOS DIAS for bar soap and GOOD MORNING for latherless shaving cream.

- LUPO for men's and boys' underwear and WOLF for clothing.

- RUST BUSTER for rust penetrating spray and BUST RUST for penetrating oil.

Examples of similar marks that have been allowed.

- LITTLE PLUMBER for liquid drain opener and LITTLE PLUMBER for advertising services.

- CATFISH BOBBERS for fish and BOBBER for restaurant services.

- GOLDEN CRUST for flour and ADOLPH'S GOLD'N CRUST for food coatings.

- DESIGNERS/FABRIC for fabric store and DAN RIVER DESIGNER FABRIC for textile fabrics.

- CROSS-OVER for bras and CROSSOVER for ladies' sportswear.

- BOTTOMS UP for ladies' and children's underwear and BOTTOMS UP for men's clothing.

- PLAYERS for men's underwear and PLAYERS for shoes.

- REPECHARGE for skin care products and SECOND CHANCE for toiletries.

- TIA MARIA for restaurant services and AUNT MARY'S for canned fruits and vegetables.

- HAUTE MODE for hair coloring and HI-FASHION for finger nail enamel.

- BEST JEWELRY for jewelry store and JEWELER'S BEST for jewelry.

- BED & BREAKFAST REGISTRY and BED & BREAKFAST INTERNATIONAL both for lodging reservations.

- COBBLER'S OUTLET for shoes and CALIFORNIA COBBLER for shoes.

- ASO QUANTUM for laboratory reagents and QUANTUM 1 for laboratory instrument.

You may notice that there does not seem to be much difference between the ones that are acceptable and the ones that are not. You have just learned what law students study for three years: There is no absolute answer in law, and any case can go either way depending on the judge, the parties, the lawyers, and the jury.

The lesson in this is: don't choose a mark that might be found to be confusingly similar to another mark, especially when the other company is bigger than yours.

CHOOSING THE CLASSIFICATION

Prior to searching your mark, you need to decide the classes into which your goods or services fall. If your goods are related, such as beers, waters and sodas, then they will all fall into the same class. However, if you plan to use one mark on such different items as perfume, clothing and jewelry, then you will have to do a separate search and file a separate application for each of these classes of goods. There will be a separate fee for each class, both when you order your search and when you file your application.

Goods and services are classified according to a system of International classifications. Prior to September 1, 1973, the United States had a separate system of classifications, but this is no longer used by the USPTO. The International classes are listed in appendix A. However, some of these are misleading. For example, computer game programs are included included in class 9, "scientific, etc." items, not class 28, "games and playthings." You can find a list of classifications for every imaginable subject at: http://www.uspto.gov/web/offices/tac/doc/gsmanual/.

Try to look several years in the future when choosing and searching your mark. If you are registering a trademark for beer, and think that you might add a line of snack crackers some time in the future, you should check that classification as well. It would be a shame to build up good will for the name for many years and not be able to use it in areas of natural expansion.

SEARCHING YOUR MARK 4

There is no requirement for a trademark search before filing your registration; but if your mark is rejected because it is too similar to an existing mark, you will not get your filing fee back. And if your mark is accepted, but violates someone's existing rights, you may get sued. Doing a trademark search lets you know if your claim to the mark will be solid and can help you avoid a lot of expense and wasted time later.

HOW THOROUGHLY TO SEARCH

The first question when beginning a trademark search is how thoroughly to search. No matter how thorough a search you do or how much you pay someone else to search, there is no guarantee that there is not someone using the mark whom you didn't miss. But if you do a thorough search and they don't turn up, you can acquire rights to the mark everywhere in the country except for the niche in which they have been using it.

The purpose of the search is to find out if anyone is already using the mark because all such prior users will have rights superior to yours, at least in the geographical area where they are doing business.

It is not enough to merely search for other registered trademarks because there may be dozens of users of the mark who have not registered it. Suppose you want to open a chain of restaurants called Bombay Tacos (curry flavored) and find that no one has registered a trademark for that name. You will probably be able to register it as a trademark (except for the geographic name problem) but if there are already restaurants named Bombay Tacos in San Francisco, Seattle and El Paso, you will never be able to use your name there. And if a restaurant in India has a website with that name, you might not be able to use a website with your trademark.

A BASIC
SEARCH

So, your goal is to do as thorough a search as your budget and time allow in consideration of how much risk you wish to take that someone is already using the name. At a minimum, you should search the following:

☞ USPTO Trademark Database

☞ Business directories

☞ Domain names

☞ Various search engines

All of these searches can be done on the internet as explained later in this chapter.

SEARCH FIRMS

To do a more thorough search, you can hire a search firm which can check all state trademarks, review the most recent federal trademark applications which are not yet listed on the USPTO website, and search numerous other directories. This can cost between $150 and $700. Two companies that provide these services are:

Government Liaison Services, Inc.   Thomson & Thomson
3030 Clarendon Blvd., Suite 209   500 Victory Road
P. O. Box 10648   North Quincy, MA 02171-1545
Arlington, VA 22210   (800) 692-8833
(800) 642-6564; (703) 524-8200

If you decide a professional search is worth the money, you should at least do a preliminary search yourself of the USPTO Database and the yellow pages listing because if it is clearly already in use, you would be wasting your money on a thorough search.

If someone is using it in the same field as you, you should choose another mark. If someone is using it or something similar in a related or unrelated field, you will need to weigh the risks and advantages of using it.

If a large company is using a similar mark in a related field, they will likely oppose your use of the mark. During the registration process your mark will be published to see if anyone opposes it and many corporations scour these listings for similar marks and file oppositions or lawsuits. (See chapter 12 on what to do if someone files an opposition.)

If the mark is being used by a small company in an unrelated field, you will probably be safe in adopting it as a trademark in your field. And if he has not registered it and you do register it, you will have the right to stop him from expanding into your field.

Even if you are already using the mark, you should have a search done before filing your application. You can register your mark if someone else is using it, but not if they have registered it for the same type of goods on which you will be using it. If others are using the mark, even in limited areas, your registration will not affect their use of the mark. If enough others are using the mark, or something similar, you may then decide it would be better to choose another mark.

With all of the thousands of companies and millions of products produced today, it is likely that something will be found somewhere that is similar to your proposed trademark. Often twenty or even fifty different marks will be found to be similar to yours. Some of these might be the exact mark used on a different type of product, or similar marks used on the exact product. You might be able to successfully register your trademark, or you might end up being sued by a large corporation. Whether you should abandon your proposed mark and come up with something different or stick with your original mark is a legal decision that is best made with the advice of an experienced trademark attorney.

However, even if you get the best trademark attorney in the country, there is no guaranty that his opinion will hold up if you are taken to court. Often, even federal judges with many years of experience are reversed by higher courts. This happened recently when Mead Data Central, owner of the Lexis legal research service, sued Toyota when it announced its new model would be called Lexus. Even though legal research services and cars are obviously in different classes of goods, and even though the marks were spelled differently, Mead argued that there would be confusion because both were sold to the same customers— lawyers. A federal judge bought the argument and forbid Toyota from using the Lexus name. Silly him. The appeals court reversed his decision and said that reasonable people would not be confused.

If your search does turn up a similar name on a similar product, it would be best not to use that name unless you have a trademark lawyer in the family who can handle your litigation. Trademark litigation costs an average of over $100,000. However, if a lot of money has already been invested in a name, or if it is important to your plans, you might want to take the chance. If the marks only have minor similarities or are in different classifications, you might not have any trouble. It is possible, too, that the other side might not contest your mark.

Also, you may contest another person's registered mark. The owner may not have filed the proper affidavit after five years to keep the registration effective, or he may have gone out of business, abandoned the mark, or otherwise acted improperly. If you wish to consider challenging a mark, you should consult an attorney who specializes in trademark law.

DOING YOUR OWN SEARCH

TRADEMARK
RECORDS
SEARCH

Up until 1999, the only ways to search the records of the United States Patent and Trademark Office were to go there, use a Trademark Depository Library, or hire a search firm to do a search. Because of the cost and difficulty, most people did a preliminary search before searching the trademark records. But now you can do a search instantly on the Internet, so that is the best place to start. The website is: http://www.uspto.gov/tmdb/index.html.

Here is what you will see:

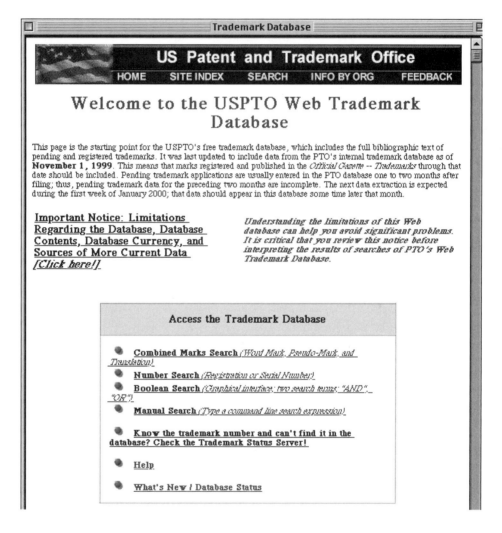

The database is updated regularly, but it is usually a few weeks behind schedule. You will see the date it is current through in the first paragraph. As a practical matter, if a mark you are considering has not been registered in the last two hundred years, it is not likely that it has been in the last few weeks, but it is possible. Especially if you are using a mark related to the latest technology. If you wish to update your mark through the latest filings, you will need to either visit the USPTO or hire a search firm to do so.

To begin your search, you will need to choose the type of search. For most purposes the first listed choice, "Combined Marks Search" is what you need. Click this and this will be the next screen:

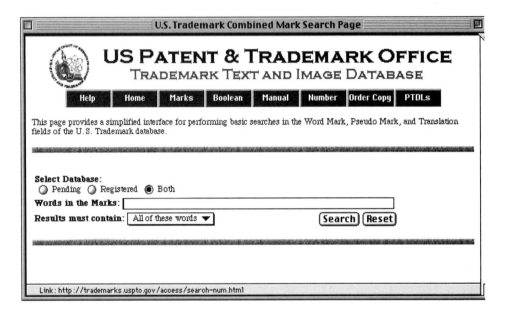

Since you will want to know about both registered marks and presently pending applications, you should be sure the "Both" button is clicked. Next, type in the words which make up your mark and click "Search." For an example, we will use a three-word mark, "NEW SWISS NAVY." On the next page is the result of the search.

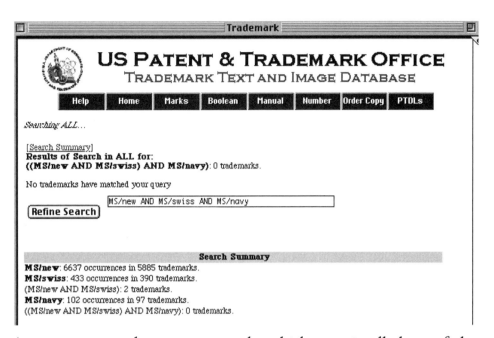

As you can see, there are no marks which contain all three of these words. However, a mark using two of them in the same category could be confusing so you should check any marks using two of the words. The search result shows that "NEW" and "SWISS" are used in two marks, but the search has not checked "SWISS" and "NAVY" in combination so we should delete the words "MS/new" and hit the "Refine Search" button. This will give the following result:

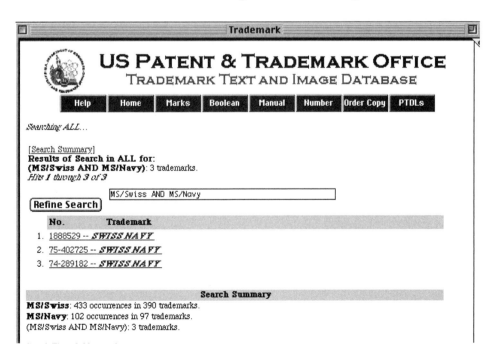

We should check each of the three using "SWISS NAVY" and the two using "NEW SWISS." Here is what you get when you click on one of the Swiss Navy marks:

At the bottom is the class in which the mark is used and the type of goods. If these are in no way related to our class, we probably do not have to worry about this mark conflicting unless it is a very famous mark.

In the first result of our search, there were the following matches:

☞ 6637 occurrences of "NEW"

☞ 433 occurrences of "SWISS"

☞ 102 occurrences of "NAVY"

While "NEW" is a common word and would not likely conflict if used in another mark, both "SWISS" and "NAVY" are somewhat unusual when used in a trademark and we should look at those registrations to see to what classes and goods they belong. If any of them are on the exact goods on which we hope to use the mark, there could be a conflict. If not, we can start checking the mark in other places.

Design Trademarks. Searching a design trademark is a lot more complicated than a word trademark and takes much more time because you need to look at nearly every mark to see if there are similarities.

To search a design, you must use a six digit numeric code representing the attributes of the design. The digits are broken into three pairs separated by periods, such as 15.05.01. The first pair of digits is the major category, for example:

01 is celestial bodies, natural phenomenon, geographical maps

02 is human beings

03 is animals

04 is supernatural or unidentifiable beings

05 is plants

06 is scenery

The list goes on to twenty-nine categories The next pairs of digits are subcategories. For example, 02.01.01 is "heads, portraits and busts of men," 02.01.02 is "shadows or silhouettes of men," 02.03.02 is "shadows or silhouettes of women."

Information on how to use the design search code is located at: http://www.uspto.gov/tmdb/dscm/index.html.

To do a design search, you need to select "Manual Search" on the first trademark search page. This will give you thirty possible fields which can be searched and the codes for searching them. For a design search the code is "DC" so you would type "DC/15.05.01" and get the following search result:

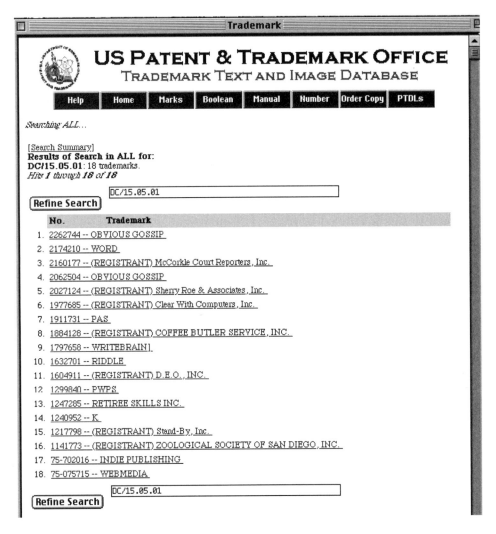

You can view each of these marks by clicking on it and the design will appear above the other registration information:

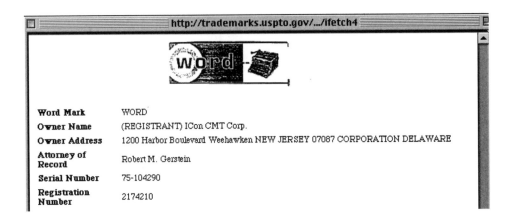

BUSINESS
DIRECTORIES

After searching the registered and pending trademarks, you should search various business directories for unregistered names and marks.

Thomas Register. The Thomas Register is a fairly comprehensive database of companies, products, and services. It is a set of large green books which is in the reference department of most large libraries. It is presently also on the Internet and can be used at no charge, although you do need to register as a user. It can be found at: http://www.thomasregister.com.

The following is the opening screen:

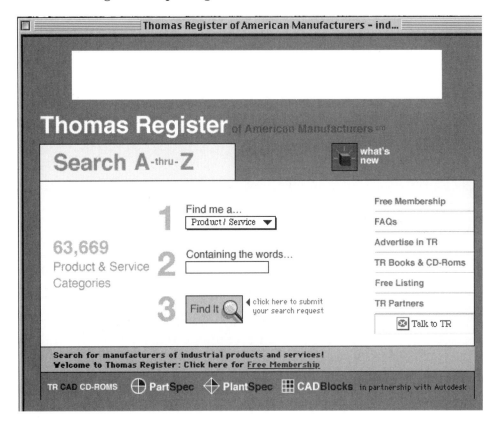

At number 1 you can choose to search companies, products/services, or brands. You should run your proposed mark through all of these to see if it is being used in any way by anyone.

Yellow pages. Major libraries have yellow page phone directories from around the country, but the easiest way to search is on the Internet. Although most search engines search by state, at least one, http://www.switchboard.com currently allows you to search the entire country at once. Since none of these search engines are one hundred percent complete, you should check a few other yellow pages directories, at least for your home state.

DOMAIN NAMES   Next, you should check to see if anyone has registered your proposed mark as a domain because if they have, it will not be available to you and they may sue you if they have also adopted it as their mark. To search domain names, you can go to: http://www.networksolutions.com/cgi-bin/whois/whois/.

SEARCH   Finally, you should do a search of your make in various Internet search
ENGINES   engines. Some of these are:

☛ http://www.altavista.com

☛ http://www.excite.com

☛ http://www.yahoo.com

This will tell you if the words you are using are used on any of the websites indexed by these engines. (Keep in mind that they only index a small fraction of all the sites and new sites appear every day.)

If you have chosen a unique word or combination of words for your mark, there most likely won't be any "hits." But if you have used common words, you may get millions. If you are searching more than one word, you should put them in quotation marks so that the search only looks for them together. Otherwise, it will find all sites that use one or both of the words anywhere on the site.

PATENT AND TRADEMARK DEPOSITORY LIBRARIES

Alabama	Auburn University Library	(205) 844-1747
	Birmingham Public Library	(205) 226-3680
Alaska	Anchorage: Z.J. Loussac Public Library	(907) 562-7323
Arizona	Tempe: Noble Library; Arizona State University	(602) 965-7010
Arkansas	Little Rock: Arkansas State Library	(501) 682-2053
California	Los Angeles Public Library	(213) 228-7220
	Sacramento: California State Library	(916) 654-0069
	San Diego Public Library	(619) 236-5813
	San Francisco Public Library	(415) 557-4500
	Sunnyvale Patent Clearinghouse	(408) 730-7290
Colorado	Denver Public Library	(303) 640-6220
Connecticut	New Haven: Science Park Library	(203) 786-5447
Delaware	Newark: University of Delaware Library	(302) 831-2965
Dist of Columbia	Howard University Libraries	(202) 806-7252
Florida	Fort Lauderdale: Broward County Main Library	(954) 357-7444
	Miami-Dade Public Library	(305) 375-2665
	Orlando: University of Central FL Libraries	(407) 823-2562
	Tampa: Tampa Campus Library, USF	(813) 974-2726
Georgia	Atlanta: Price Gilbert Mem. Library, GIT	(404) 894-4508
Hawaii	Honolulu: Hawaii State Public Library System	(808) 586-3477
Idaho	Moscow: University of Idaho Library	(208) 885-6235
Illinois	Chicago Public Library	(312) 747-4450
	Springfield: Illinois State Library	(217) 782-5659
Indiana	Indianapolis-Marion County Public Library	(317) 269-1741
	West Lafayette: Purdue University Libraries	(317) 494-2873
Iowa	Des Moines: State Library of Iowa	(515) 281-4118
Kansas	Wichita: Ablah Library, Wichita State University	(316) 574-1611
Kentucky	Louisville Free Public Library	(502) 561-5652
Louisiana	Baton Rouge: Troy H. Middleton Library, LSU	(504) 388-2570
Maryland	College Park: Engineering & Physical Sciences Library, University of Maryland	(301) 405-9157
Massachusetts	Amherst: Physical Sciences Library, Univ. of MA	(413) 545-1370
	Boston Public Library	(617) 536-5400 ext. 265
Michigan	Ann Arbor: Engineering Transportation Library, University of Michigan	(313) 647-5735
	Big Rapids: Abigail S. Timme Library Ferris State University	(616) 592-3602
	Detroit Public Library	(313) 833-3379
Minnesota	Minneapolis Public Library & Information Center	(612) 372-6570
Mississippi	Jackson: Mississippi Library Commission	(601) 359-1036
Missouri	Kansas City: Linda Hall Library	(816) 363-4600
	St. Louis Public Library	(314) 241-2288 ext. 390
Montana	Butte: Montana College of Mineral Science & Technology Library	(406) 496-4281

Nebraska	Lincoln: Engineering Library Univ. of Nebraska	(402) 472-3411
Nevada	Reno: University of Nevada/Reno Library	(702) 784-6579 ext. 257
New Hampshire	Durham: University of New Hampshire Library	(603) 271-2239
New Jersey	Newark Public Library	(201) 733-7782
	Piscataway: Library of Science & Medicine, Rutgers University	(908) 445-2895
New Mexico	Albuquerque: Univ. of NM General Library	(505) 277-4412
New York	Albany: New York State Library	(518) 474-5355
	Buffalo and Erie County Public Library	(716) 858-7101
	New York Public Library (the Research Library)	(212) 592-7000
North Carolina	Raleigh: D.H. Hill Library, NC State University	(919) 515-3280
North Dakota	Grand Forks: Chester Fritz Library, Univ. of ND	(701) 777-4888
Ohio	Akron-Summit County Public Library	(330) 643-9075
	Cincinnati and Hamilton County, Public Library	(513) 369-6936
	Cleveland Public Library	(216) 623-2870
	Columbus: Ohio State University Library	(614) 292-6175
	Toledo/Lucas County Public Library	(419) 259-5212
Oklahoma	Stillwater: Oklahoma State Univ. Library	(405) 744-7086
Oregon	Salem: Oregon State Library	(503) 768-6786
Pennsylvania	Philadelphia, Free Library	(215) 686-5331
	Pittsburgh, Carnegie Library	(412) 622-3138
	University Park: Pattee Library, PSU	(814) 865-4861
Puerto Rico	Mayaguez: General library, University of Puerto Rico	(787) 832-4040 ext. 4359
Rhode Island	Providence Public Library	(401) 455-8027
South Carolina	Clemson University Libraries	(864) 656-3024
Tennessee	Memphis & Shelby County Public Library and Information Center	(901) 725-8877
	Nashville: Stevenson Science Library Vanderbilt University	(615) 322-2717
Texas	Austin: McKinney Engineering Library Univ. of Texas at Austin	(512) 495-4500
	College Station: Sterling C. Evans Library, Texas A & M University	(409) 845-3826
	Dallas Public Library	(214) 670-1468
	Houston: The Fondren Library, Rice Univ.	(713) 527-8101 ext. 2587
	Lubbock: Texas Tech Library	Not yet operational
Utah	Salt Lake City: Marriott Library, Univ. of UT	(801) 581-8394
Vermont	Burlington: Bailey/Howe Library, University of Vermont	Not yet operational
Virginia	Richmond: James Branch Cabell Library Virginia Commonwealth Univ.	(804) 828-1104
Washington	Seattle: Engineering Library, Univ. of WA	(206) 543-0740
West Virginia	Morgantown: Evansdale Library, WV Univ.	(304) 293-2510 ext. 113
Wisconsin	Madison: Kurt F. Wendt Library, Univ. of WI	(608) 262-6845
	Milwaukee Public Library	(414) 286-3051
Wyoming	Casper: Natrona County Public Library	(307) 237-4935

PREPARING YOUR DRAWING 5

As of October 31, 1999, a drawing is not a technical requirement of a trademark application *if* the mark is included on the application form. However, including a drawing is "encouraged" by the Trademark Office, and if the mark is artwork rather than a word, then as a practical matter, it should be on a separate drawing page.

In the past, the drawing had to strictly comply with the *Rules of Practice in Trademark Cases*, 37 CFR § 2.52 as to type of paper, margins, etc. But with the passage of the Trademark Law Treaty Implementation Act of 1998, the rules have been relaxed. Note that the changes in the rules have not yet been included in the *Trademark Manual of Examining Procedure* available at press time.

There are two types of drawings. If a mark consists only of words and punctuation, the drawing is merely typed in capital letters. If the mark is a picture or stylized letters, the drawing is considered a *special form drawing*. If the mark is a scent, a sound, or otherwise non-visual, no drawing is required, but a detailed written description must be included.

Most marks are registered as typed words. That way, they are protected in any style. Only if you wish to use an artistic logo should you create a special form drawing.

Although the rules have been relaxed to accept drawings which might not comply with all of the rules, it is best to follow the rules as closely as possible to avoid delay in the examination process. The following are the requirements under Rule 2.52(b):

(b) Recommended format for special form drawings--

(1) Type of paper and ink. The drawing should be on a piece of non-shiny, white paper that is separate from the application. Black ink should be used to depict the mark.

(2) Size of paper and size of mark. The drawing should be on paper that is 8 to 8-1/2 inches (20.3 to 21.6 cm.) wide and 11 to 11.69 inches (27.9 to 29.7 cm.) long. One of the shorter sides of the sheet should be regarded as its top edge. The drawing should be between 2.5 inches (6.1 cm.) and 4 inches (10.3 cm.) high and/or wide. There should be at least a 1 inch (2.5 cm.) margin between the drawing and the edges of the paper, and at least a 1 inch (2.5 cm.) margin between the drawing and the heading.

(3) Heading. Across the top of the drawing, beginning one inch (2.5 cm.) from the top edge, the applicant should type the following: Applicant's name; applicant's address; the goods or services recited in the application, or a typical item of the goods or services if numerous items are recited in the application; the date of first use of the mark and first use of the mark in commerce in an application under section 1(a) of the Act; the priority filing date of the relevant foreign application in an application claiming the benefit of a prior foreign application under section 44(d) of the Act. If the information in the heading is lengthy, the heading may continue onto a second page, but the mark should be depicted on the first page.

If you use a typed drawing for your mark, you must only use characters that can be represented by pica or elite type. You can use the symbols, . ? " - ; ( ) % $ @ + , ! ' : / & # * = [and]. You cannot use symbols such as the degree symbol (°), subscripts, underlining or exponents. To do anything like this, you must use a stylized drawing.

If the mark you wish to register is stylized or is artwork, you will need to provide a drawing. While the requirements were liberalized in late 1999, it is still important that the drawing be clear and black. Here are the new requirements under Rule 2.52(a)(2):

(2) Special form drawing. A special form drawing is required if the mark has a two or three-dimensional design; or color; or words, letters, or numbers in a particular style of lettering; or unusual forms of punctuation.

(i) Special form drawings must be made with a pen or by a process that will provide high definition when copied. A photolithographic, printer's proof copy, or other high quality reproduction of the mark may be used. Every line and letter, including lines used for shading, must be black. All lines must be clean, sharp, and solid, and must not be fine or crowded. Gray tones or tints may not be used for surface shading or any other purpose.

(ii) If necessary to adequately depict the commercial impression of the mark, the applicant may be required to submit a drawing that shows the placement of the mark by surrounding the mark with a proportionately accurate broken-line representation of the particular goods, packaging, or advertising on which the mark appears. The applicant must also use broken lines to show any other matter not claimed as part of the mark. For any drawing using broken lines to indicate placement of the mark, or matter not claimed as part of the mark, the applicant must include in the body of the application a written description of the mark and explain the purpose of the broken lines.

(iii) If the mark has three-dimensional features, the applicant must submit a drawing that depicts a single rendition of the mark, and the applicant must include a description of the mark indicating that the mark is three-dimensional.

(iv) If the mark has motion, the applicant may submit a drawing that depicts a single point in the movement, or the applicant may submit a square drawing that contains up to five freeze frames showing various points in the movement, whichever best depicts the commercial impression of the mark. The applicant must also submit a written description of the mark.

(v) If the mark has color, the applicant may claim that all or part of the mark consists of one or more colors. To claim color, the applicant must submit a statement explaining where the color or colors appear in the mark and the nature of the color(s).

(vi) If a drawing cannot adequately depict all significant features of the mark, the applicant must also submit a written description of the mark.

At present the USPTO deleted the color lining chart which was previously required, but it does not yet have the capability to scan marks in color. They will no longer reject drawings which are in color, and they will still accept marks using the color lining codes which are shown here:

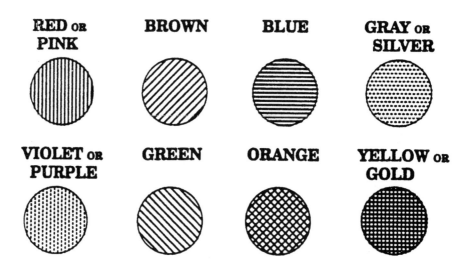

If the drawing contains these lines for color, a statement must also be included that explains which colors are designated by the lining. A commercial color designation system (such as PMS) can be used to identify the color and color photographs are also allowed.

If the mark is a drawing that is shaded, but not for color, then a statement should be included which indicates that the shading lines do not represent color (for example, "The lining shown in the mark is a feature of the mark and is not intended to indicate color").

If the mark is shaded for color, but you are not claiming the color to be part of your trademark, then a statement should be included which makes clear that you are not claiming rights to the color as part of the mark. For example, "but no claim is made to color," or "but color is not a feature of the mark." Unless you are sure that you will never want to use the mark in another color, you should not claim color as a feature of the mark.

If your stylized mark contains words that cannot be exclusively claimed, such as "Sunshine Georgia Peaches," you should disclaim rights to those extra words (for example, "No right is claimed to the exclusive right to use 'Georgia' or 'Peaches' apart from the mark as shown"). If your mark was not stylized, only the word "Sunshine" would be typed on your drawing.

If the mark is a sound or a scent, the drawing page should indicate "NO DRAWING" where the mark would otherwise appear.

A typed drawing may be folded as long as the fold does not run through the mark. A special-form drawing or any drawing on bristol board should not be folded.

Sample typed and stylized drawings are on the next pages.

Sample Typed Drawing

APPLICANT'S NAME: Sphinx International, Inc.

APPLICANT'S ADDRESS: 1725 Clw/Largo Rd.
                     Clearwater FL 33756

GOODS: Law books and legal forms

DATE OF FIRST USE: July 20, 1999

DATE OF FIRST USE IN COMMERCE: July 20, 1999

SPHINX

Sample Stylized Drawing

APPLICANT'S NAME: On call, Inc.

APPLICANT'S ADDRESS: 1122 Ridge Rd. Anywhere, MT 01234

SERVICES: Telephone answering service, telephone books

FIRST USE: Answering services (Class 38) June 20, 1999
   Telephone books (Class 16) May 6, 1998

FIRST USE IN COMMERCE:
   Answering services (Class 38) June 20, 1999
   Telephone books (Class 16) January 29, 1999

DESIGN: A telephone

CHOOSING THE PROCEDURE 6

A few years ago, a mark had to be in actual use in interstate or international commerce before an application for registration could be filed. This was a problem for large companies that spent millions of dollars to introduce new products because that money would be wasted if the mark was later rejected. To avoid this problem, companies would make a *token use* of the mark and then file the application. Token use usually meant shipping a sample product to a friend or relative in another state.

On November 17, 1989, the first major overhaul of the trademark law in forty years took effect. The biggest change was the provision allowing *intent to use* applications. This procedure allows an application for a trademark to be filed by a person who has a *bona fide intent to use* a mark.

The person then has six months to start using the mark after the application is filed. This six month period may be extended an additional six months without giving a reason, and an additional four more six-month extensions may be granted if good reason is given. Therefore, a person has a total of three years in which to make use of the mark. The fee for each six-month extension is $100.

Chapter 7 explains the procedure for filing an Intent to Use trademark application. If it will be a while before you can start legitimately using your mark, you should file an application as explained in chapter 7.

If you can quickly begin using your mark, or are already using it, you should skip chapter 7 and go right to chapters 8 and 9 which explain proper use of the mark and the type of specimens required. You can then file an application for a mark in use as explained in chapter 10.

As explained in chapter 8, token use is no longer allowed. The mark must be used in an actual transaction either after an Intent to Use application has been filed, or before a Use application is filed.

If you claim use but do have not had a bona fide transaction you will most likely still have your application accepted since the USPTO does not investigate use. But at a future time, if a competitor challenges your mark, you may lose it if you cannot prove legitimate use.

Your use of the mark the first time you use it does not have to be the same format on which you plan to use it in the future. If you plan to have some fancy die-cut labels prepared to affix your mark to the goods, but they will not be delivered for several weeks, you can prepare interim labels on your laser printer. The important factor is that the mark you use is the same you are registering and that you affix it to your goods (or use it in conjunction with your services) to identify them as your own.

So, if you can possibly make legitimate use of your mark immediately, you should go on to chapter 8. If you cannot make use of your mark yet, continue on to chapter 7.

FILING AN INTENT TO USE APPLICATION 7

This chapter explains how to file your application for a trademark you have not yet used in interstate commerce, but for which you have a "bona fide intention" to use.

Because this procedure is relatively new, there is not a lot of legal guidance as to what constitutes proof of such an intention. However, in 1993, the Trademark Trial and Appeal Board made clear what is *not* a bona fide intention. In the case *Commodore Electronics Ltd. v. CBM Kabushiki Kaisha*, Commodore challenged CBM's application and asked CBM to produce proof that it intended to use the mark in all of the classes claimed. The Board ruled that a lack of any *documentation* proving an intent to use the mark is sufficient proof that there was no valid intention.

What type of documentation is necessary? A narrative of your plans dated and notarized prior to filing the application would be better than nothing. But the best documentation would be actual plans, memos, and minutes of meetings explaining in detail the plans to use the mark.

The procedure for registering a mark that has not yet been used consists of two steps. The first step is to file an application along with a drawing of the mark and the filing fee at the Patent and Trademark Office. Then, after you have used the mark, you must send in a specimen of its use along with either an Amendment to Allege Use or a Statement of Use and an additional filing fee.

For an application to be complete, it must contain the following elements:

1. The name and address of the applicant.

2. The name and address of a person to whom communications can be addressed. (This is usually the applicant or his attorney.)

3. A clear drawing of the mark.

4. A listing of the goods or services on which the mark will be used.

5. A claim that the applicant has a bona fide intent to use the mark in commerce.

6. A verification or declaration paragraph.

7. A filing fee.

The requirements were liberalized in 1999 to allow some elements to be sent in later. The ones absolutely needed to begin the filing are 1, 2, 3, 4, and 7. The others will be requested if not included initially. However, keep in mind that your application will be processed more quickly if all elements are included on the correct form.

PREPARING THE APPLICATION

The easiest way to apply is to use the Patent and Trademark Office forms that are included in this book. It is also possible to type the information on blank 8-1/2" x 11" paper but processing will be slower since it cannot be scanned into the USPTO computers. To see if there is a more recent version of the form, you can check the United States Patent and Trademark Office website, and if necessary, download the form at: http://www.uspto.gov/web/forms/.

The USPTO now supplies five pages of line-by-line instructions for the form which are included in appendix C following the form. We have also included a sample filled-in form in appendix B. Some important tips are:

☞ The mark on the application should be exactly the same as the drawing you submit.

☞ If you are not sure of the classification of the goods, you can leave it blank and the PTO will fill in the correct class.

☞ If the goods you are going to use it on fall into more than one class, you must pay a $325 fee for each class.

☞ The applicant must be the owner of the mark who actually controls the nature and quality of the goods sold.

☞ You should use care in describing the goods and services. The more narrowly you describe your claim, the more easily it will be approved, but the smaller your protection. For example, if you sell a blood test kit, you might want it to be considered a "medical test kit" so that no one else could use the mark on any type of medical test kit. If you claim the mark just for a "blood test kit," then you will be less likely to conflict with existing marks in the medical field that are similar, but others might be able to use a similar mark on a different type of medical test kit.

☞ Under "Basis for Application," you should ignore section 1(a) (that is for a mark already in use) and complete section 1(b).

SUBMISSION

A completed application consists of the application itself, the drawing, and one application fee for each class in which the mark is to be registered. (As of January 10, 2000, the fee was $325) If you send the wrong fee it will be sent back, which will delay your filing date so you may want to confirm the fee by calling 703-308-4357. The address is:

Assistant Commissioner for Trademarks
2900 Crystal Drive
Arlington, VA 22202-3513

Your application package should be sent in by certified mail, return receipt requested. If you send in your package by Express Mail and properly fill in the Express Mail number on your application, the date you mail it will be considered your filing date (unless that day is Saturday, Sunday or a federal holiday in the District of Columbia). The necessary wording is on the application forms in this book. If you type your application, you must use this same wording.

After the application has been sent in, it will be reviewed for errors and published in the *Official Gazette*. These procedures are explained in chapters 10 and 11 of this book.

Step Two in Intent to Use Applications

Prior to final registration of a mark, an Intent to Use applicant must use the mark in commerce and must file either an Amendment to Allege Use or a Statement of Use along with a specimen and an additional filing fee. These used to be two separate forms, but now one form, the Allegation of Use for Intent-to-Use Application, covers both situations. This form will be considered an Amendment to Allege Use or a Statement of Use depending upon when it is filed. A description of the proper type of use is explained in chapter 7, and an explanation of the proper specimens is included in chapter 8.

AMENDMENT TO ALLEGE USE

The Amendment to Allege Use may be filed any time after filing the application and before the approval for publication in the Official Gazette. So, if you begin using the mark before it has been approved for publication, you should file an Amendment to Allege Use. A complete Amendment to Allege Use must contain the following:

☛ A verified statement that the applicant is believed to be the owner of the mark and that the mark is in use in commerce; the type of commerce; the date of the applicant's first use of the mark and first use of the mark in commerce; those goods or services specified in the application on or in connection with which the mark is in use in commerce; and the mode or manner in which the mark is used.

☛ A specimen of the mark as used in commerce.

☛ The filing fee of $100.

An Allegation of Use for Intent-to-Use Application form is included in this book as form 7. As an alternative to using the included form, the information listed above may be typed on white 8-1/2" x 11" paper and be titled "Amendment to Allege Use" at the top.

BLACKOUT PERIOD A blackout period exists between the approval for publication and the issuance of the Notice of Allowance. During which neither the Amendment to Allege Use nor the Statement of Use may be filed.

STATEMENT OF USE The Statement of Use may be filed after issuance of a Notice of Allowance. It must be filed within six months, unless an extension has been requested. This notice is issued after publication of the mark if there were no oppositions, or after the applicant has dealt with any oppositions. The Statement of Use must contain the following:

☛ A verified statement that the applicant is believed to be the owner of the mark and that the mark is in use in commerce; the type of commerce, specifying the date of the applicant's first use of the mark and first use of the mark in commerce; those goods or services specified in the notice of allowance on or in connection with which the mark is in use in commerce; and the mode or manner in which the mark is used.

☛ A specimen of the mark as used in commerce.

☛ The filing fee of $100.

A form Allegation of Use for Intent-to-Use Application is included in this book as form 7. As an alternative to using the included form, the information listed above may be typed on white 8-1/2" x 11" paper and be titled "Statement of Use" at the top.

SIX-MONTH
EXTENSION

A six-month extension can be added to the initial six-month period for filing the Statement of Use. To obtain the extension, the applicant must file a Request For Extension during the initial six-month period. This form is included in this book as form 8. Instead of using this form, the request may be typed on white 8-1/2" x 11" paper. It must include the following:

☛ A verified statement signed by the applicant that the applicant has a continued bona fide intention to use the mark in commerce and a specification on which goods or services the applicant intends to use the mark.

☛ The fee of $150.

Up to four additional six-month extensions can be granted if the applicant can show good cause why it has not been possible yet to use the mark in commerce. To obtain the extension, the applicant must file a Request for Extension (as explained above) and include a statement showing a good reason for why the extension is necessary.

REQUEST TO
DIVIDE

If you have applied for a trademark on several classes of goods or services and only used it on some of the classes, you may divide your application and ask that the trademark be issued for the classes in use. Then you can get the trademark for the other classes at a later time after you have used the mark on those classes.

To do so you must file a Request to Divide which is included in this book as form 9.

A Request to Divide may be filed where the mark has been used on some of the intended goods, but not all. This way, a registration can be granted for the mark on some goods even if it will not be used on others until much later. A form for this purpose is included in this book.

SECURITY
INTEREST

After filing an intent-to-use application, you should not include it on any assignments of security interest in your company's assets (for example to secure a bank loan). In one case a trademark registration was held void because the applicant had granted a security interest in it to its bank. *Clorox Co. v. Chemical Bank*, 40 USPQ 2d (TTAB 1996).

USING YOUR MARK 8

A trademark registration cannot be finalized until the mark has been *used in commerce*. Back when the law required a mark to be used before an application could be filed, many applicants just made some token use of their mark, such as mailing a copy to a friend, in order to have a first date of use of the mark.

Because Intent to Use applications are now allowed, the requirements for actual use of the mark have been made more strict. Token use is no longer acceptable. Be sure to follow these rules in making use of your mark.

IN COMMERCE   One of the basic requirements for trademark registration is that the mark must have been used in commerce. *In commerce* is a legal term that means the mark must have been used in either interstate commerce or commerce with a foreign country. The specifics of the requirement are as follows:

1. The mark must have been affixed to the goods or the packaging, labels, or on tags on the goods. The mark should not be handwritten. It should appear there as an intention to make serious use of the mark as a trademark.

2. The product should be sold to an unrelated party at a normal price, and the product should not be returned nor the payment refunded. The sale should be in the ordinary course of business and not just

for the purpose of using the mark. It is best to have some sort of invoice or other written documentation of the sale.

One court ruled that shipping a drug for clinical testing was sufficient use in commerce, but it is much safer to rely on an actual sale of the goods. *G.D.Searle v. Nutrapharm*, No. 98-6890, DCNY Oct 29, 1999.

SERVICE MARKS   When using service marks, such as for a restaurant where no goods are shipped across state lines, the mark must appear on advertising that is sent across state lines. The advertising should be for services that are currently available for sale, not merely an announcement of future availability.

COLLECTIVE MARK   To use a collective mark, such as the name of a club, the mark should be used on membership forms, stationery, etc.

CERTIFICATION
MARK   A certification mark must be used by a party who is not the owner of the mark but whose products or services comply with the standards set for the mark. The owner of the mark must set definite standards for use of the mark and must be sure that users of the mark comply with the standards.

Prior to registration of a mark, you may claim ownership of a mark by using the symbol ™ next to your mark. If you are using a service mark, the symbol consists of the letters ℠. Even if you never register your mark with the Patent and Trademark Office, you can use these symbols to indicate your claim to ownership of a mark that you are using as a trademark on your goods or services.

If the mark has been registered in a foreign country, or if an application for registration has been filed in a foreign country, then use in commerce in the United States is not necessary. If relying on a foreign *application*, your U.S. application must be filed within six months of the foreign filing. If relying on a foreign *registration*, a certification or a certified copy of the foreign registration must be included.

PREPARING YOUR SPECIMEN 9

In 1999, the number of specimens required by the USPTO was reduced from three to just one. This is sent in either with the Present Use Application, or, if an Intent to Use Application is filed, it is sent with the Amendment to Allege Use or the Statement of Use.

The specimen should be an actual example of how the mark is used to identify the goods or services. It may be a tag, label, container, display, or some similar use of the mark that can be arranged flat and is not larger than 8-1/2 by 11 inches. Since the specimen needs to be kept in your file at the Trademark Office, three dimensional or bulky specimens are not acceptable. If specimens complying with the above rules cannot be supplied (for example, if the mark is a metal ornament on an automobile or the shape of a bottle), a photograph of the mark may be submitted. The photograph must also be no larger than 8-1/2 by 11 inches.

Advertising materials are not acceptable as specimens for goods. Neither are instruction manuals, invoices, business stationery, nor bags used in a store at checkout. You must send actual labels or tags that are *affixed* to the goods. However, for services, such as a restaurant, advertising materials are acceptable. For marks that only appear on a video or television screen, a photograph of the screen with the mark displayed is acceptable.

If the trademark drawing is coded for specific colors on the mark, then the specimen must be in the correct colors.

The specimen must match the mark sought to be registered. If your drawing was typed, you can use any font style on your specimens, but the spelling should be identical.

If some material printed on the specimen is not in English, the examining attorney may require a translation of it "to permit a proper examination."

Since the symbol ® can only be used on a mark that has been registered, it should not be used on a specimen of marks that you are in the process of applying for registration. You may, however, use the ™ symbol.

FILING AN APPLICATION FOR A MARK IN USE

10

If you have already used your mark in commerce, or if you can easily use your mark, then filing an Application for a Mark in Use is the best way to register it. It is easier and cheaper than filing an Intent to Use Application.

For a mark that has already been used in commerce, the following items are required for a complete application:

☛ the written application;

☛ the drawing of the mark;

☛ a specimen showing actual use of the mark; and

☛ the filing fee.

You should also include a self-addressed stamped envelope so that you will receive a receipt for your application.

PREPARING THE APPLICATION

For an application to be complete it should contain the following elements:

☛ The name and address of the applicant.

☛ The name and address of a person to whom communications can be addressed. (This is usually the applicant or the applicant's attorney.)

- ☞ The citizenship of the applicant.

- ☞ The classes in which the applicant wishes to register the mark. (See appendix A for the listings of classes.)

- ☞ A statement that the applicant has adopted and is using the mark.

- ☞ A drawing of the mark.

- ☞ Identification of the goods or services.

- ☞ A request that the mark be registered under the Trademark Act.

- ☞ The date on which the mark was first used in commerce.

- ☞ The type of commerce in which the mark was used (that is, interstate commerce or international commerce).

- ☞ The manner in which the mark is affixed to the goods or used to promote the services.

- ☞ If the applicant is represented by an attorney, then a statement should be included appointing the attorney to prosecute the application.

- ☞ An oath that the statements are true. This can be notarized or in the form of a declaration. The forms in this book use the declaration rather than notary.

The easiest way to file an Application for a Mark in Use is to use one of the Patent and Trademark Office forms that are included in this book. These were redesigned in 1999 to allow the material to be scanned into the USPTO computers. It includes four pages of line-by-line instructions.

To show you how to fill in the forms, sample filled-in application forms are included in appendix B of this book. In addition, be sure to follow these instructions which are included in appendix C following the form. We have also included a sample filled-in form in appendix B. Some important tips are:

- ☞ The mark in your application should be the same as the drawing you submit.

- ☞ You are not required to fill in the "Class No."—the USPTO will fill it in for you. However, you will already know the class from doing your search, so you might as well record it in case there is any chance for confusion. If you wish to register in more than one class, list them all here and be sure to pay a separate filing fee for each class.

☞ The applicant must be the owner of the mark who actually controls the nature and quality of the goods sold.

☞ You should use care in describing the goods and services. The more narrowly you describe your claim, the more easily it will be approved, but the smaller your protection. For example, if you sell a blood test kit, you might want it to be considered a "medical test kit" so that no one else could use the mark on any type of medical test kit. If you claim the mark just for a "blood test kit," then you will be less likely to conflict with existing marks in the medical field that are similar, but other might be able to use a similar mark on a different type of medical test kit.

☞ Under "Basis for Application," you should complete section 1(a) but not section 1(b) (which is for intent to use applicants).

☞ The date of first use in commerce and the date of first use anywhere should be exact dates. If only a month and year are given, the last day of the month is presumed. If only a year is given, the last day of the year is presumed.

FOREIGN APPLICATION OR REGISTRATION
Applicants who rely on a foreign application or registration do not need specimens; do not need to give the dates of use; nor the manner of affixing the mark to the goods. Applicants relying on a foreign application must specify their priority date in the heading of their application.

FOREIGN APPLICANTS
Foreign applicants must have a *domestic representative*. This is a resident of the United States on whom notices of process can be served regarding the application. The domestic representative will also receive official correspondence unless the applicant is also represented by an attorney in the United States. Designation of a domestic representative can be accomplished with form 14, or the same wording can be added to one of the application forms or to a typed application.

SUBMISSION

A completed application consists of the application itself, the drawing, a specimen, and one application fee for each class in which the mark is to be registered. (As this book was going to press, the fee was $325.) You should write or call the Patent and Trademark Office to confirm the cost of the fee before filing. If you send the wrong fee, it will be sent back, which will delay your filing date. The phone number is 703-308-4357. The mailing address is:

> Assistant Commissioner for Trademarks
> 2900 Crystal Drive
> Arlington, VA 22202-3513

Your application package should be sent in by certified mail, return receipt requested. If you send in your package by Express Mail and properly fill in the Express Mail number on your application, the date you mail it will be considered your filing date (unless that day is Saturday, Sunday, or a Federal holiday in the District of Columbia). The necessary wording is on the application forms in this book. If you type your application, you must use this same wording.

After the application has been sent in, it will be reviewed for errors and published in the *Official Gazette* as explained in the next chapter.

WHAT HAPPENS AFTER FILING 11

After your application has been mailed and the receipt returned, one of several things can happen. The least likely is that your trademark will be routinely registered. More likely, you will get a letter or phone call requesting more information or requiring changes in your application.

INCOMPLETE APPLICATION

One thing that may happen is that you may get your whole packet back along with a letter stating that your application is incomplete. There is a form letter with a checklist of the possible reasons it may be incomplete. This checklist is included below. You should look it over to see why your application might be returned. However, the 1999 changes in USPTO policy will mean that fewer applications will be returned and more applicants will be asked for additional information.

1. The applicant has not been identified by name.

2. Applicant has not provided an address to which communication can be directed. (How did they return your application?)

3. The drawing requirements have not been met.

4. No drawing has been submitted.

5. The drawing heading is incomplete. A complete heading includes applicant's name, post office address, dates of first use of the mark both anywhere and in commerce, and the goods recited on the application.

6. The goods or services in connection with which the mark is used have not been identified.

7. The requirements for specimens have not been met. [Some of these reasons do not apply to Intent to Use applications.]

8. The materials submitted as specimens are not acceptable because they are merely reproductions of the drawing.

9. No specimen has been submitted.

10. Your attempted use of the mark has been defective in one of the following ways:

    • The date of use in commerce must be specified and must be prior to the filing date, or

    • To use a foreign application for a basis for use, the U.S. application must be filed within six months of filing the foreign application and must state the country of application, or

    • To use a foreign registration for a basis of use, a certification or certified copy of the registration must be submitted.

11. The correct fee was not included.

12. No fee was included.

13. Your check was unsigned.

14. Your check was not drawn on a U.S. bank.

15. The specimens were too bulky.

After the application has been properly filed and accepted, you will get a small receipt indicating the filing. This should come in the self-addressed, stamped envelope that you included with your application.

USPTO
QUESTION

A second thing that can happen is that you can receive a letter or phone call from the examining attorney at the Patent and Trademark Office regarding some legal or factual problem with the application. Usually, the examiner will tell you what the problem is with your application and what needs to be done to correct it. Pay close attention to his or her instructions. Do exactly what he or she requests. Read each paragraph of the letter and be sure to follow it or answer it exactly. You are given the attorney's phone number so that you can call to ask questions to determine what would be acceptable.

In some cases, you might want to negotiate with the examiner. If you are asked to limit your mark to a very narrow class of goods but are using it or plan to use it in a broader class, ask if you can broaden the definition. For example, if you used the mark on a book and the examiner wants to limit your mark to "books in the field of political science" you could argue that the book also includes economics and sociology so that you wish the mark to cover "books in the field of political science, sociology and economics." This is important because your mark would be more valuable if it protected you in a broader field.

RESPONSE TO
USPTO

Response to the examiner's letter may be an explanation or submission of additional materials, or it may require an amendment of the application. If an amendment is required, it should be typed on 8-1/2" x 11" paper with the heading "Amendment" and refer to the original application. It should explain exactly which parts of the application are to be changed in which way.

For example, an amendment might read:

In response to Office Action of January 29, 1999, please amend the application referred to above as follows:

On page 1, line 2, insert the word -plumbing- before the word "books."

On page 1, line 9, replace the words "placing on" with the words -gluing labels onto-.

Note that the communication from the Patent and Trademark Office is referred to as *Office Action,* and the wording that is contained in the original application is put in quotation marks while new wording is surrounded by dashes.

Normally, you will be given six months to send a letter or amendment. There is no reason to wait this long. You can usually clear the problems up with a letter that should be sent out immediately. If you do address each issue brought up by the examiner, but have not answered them sufficiently, you will usually be given an additional time period to make further corrections. If you fail to answer any issue, then your application will be considered abandoned.

SERIOUS PROBLEMS
If the problems with your application seem serious, you should consult the *Trademark Manual of Examining Procedure* as explained on page 28. You may also wish to consult the Rules of Practice in Trademark Cases contained in 37 CFR Ch.1 and the Trademark Act contained in Title 15, Chapter 22, United States Code, both of which should be in your local law library or can be found on the USPTO's website.

If you are not an attorney and are processing your own application, you may want to consider using the services of an attorney to insure completion of your registration if the problems seem to be serious. If your application is rejected a second time, it may be final and an appeal would be much more expensive. If you do decide to consult an attorney, you should be sure to choose one who has experience with trademarks. When making an appointment, you should explain your situation and ask if the attorney would be willing to draft your amendments or take over your case.

ACCEPTANCE
Once your application has been accepted as correctly filed, your mark is ready to be published. If you have filed an Intent to Use application, you cannot file an Amendment to Allege Use after your mark has been approved for publication. You must wait until you receive the Notice of Allowance and then file a Statement of Use.

STATUS
It usually takes several months before you hear from the Patent and Trademark Office. But you can check on the status of your application in the meantime in any of three ways:

☞ You can check the status on the USPTO website: http://tarr.uspto.gov/.

☞ You can call their automatic Trademark Reporting And Monitoring system (TRAM) at 703-305-8747 between 6:30 A.M. and midnight Eastern Time.

☞ You can talk to a live person at 703-308-9400.

If you do not hear from the USPTO in six months, you should check on the status to be sure that some correspondence has not been lost.

PUBLICATION Once your application is acceptable, your mark will be published in the *Official Gazette* as explained in the next chapter.

PUBLICATION AND OPPOSITION 12

PUBLICATION OF YOUR MARK

Once your trademark application has been reviewed by the trademark office's examining attorney and found acceptable, the next step is publication of your mark in the *Official Gazette of the United States Patent and Trademark Office*. This publication comes out each week and lists new trademarks, trademarks renewed, cancelled, etc. A sample page from the *Official Gazette* is shown on page 98. The price for a single copy is about $12 and you will be given instructions on how to order the issue containing your mark. The purpose of the publication is to let the world know of your application and to see if anyone opposes your registration. There are companies that are hired to read the Gazette and inform businesses if someone is attempting to register a mark similar to their own.

OPPOSITION TO YOUR MARK

Someone opposing your mark has two choices of action. He or she can file a Letter of Protest with the Director of the Trademark Examining Operation, or he or she can file an *opposition*. Handling an opposition is beyond the scope of this book. Anyone who has had an opposition filed against him or her should consult an attorney or begin some serious research of the *Rules of Practice in Trademark Cases*, § 2.101 to § 2.148, or a treatise on the subject such as *Trademark Law and Practice* by Edward C. Vandenburgh. However, a brief explanation of the procedures follows:

LETTER OF
PROTEST

A *letter of protest* would point out some reason your mark should not be registered, such as another registered mark that is confusingly similar or evidence that you are attempting to register a generic mark. In this way, a person hopes to point out a serious enough problem with your application that it will be rejected by the Trademark Office.

OPPOSITION

An *opposition* is an action similar to a lawsuit between the parties that is brought before the Trademark Trial and Appeal Board. The proceeding is conducted just like a trial in Federal District Court and is governed by the Federal Rules of Civil Procedure.

NOTICE OF
OPPOSITION

A *notice of opposition* may come from a large law firm and include a threat of a separate lawsuit for infringement in Federal District Court. In some cases, this may be a serious threat and you may be advised to immediately discontinue use of the mark. This is what the law firms want. But often the letter is just a bluff. Some companies don't want anyone to use any name which in any way resembles their mark. However, they have no right to stop the use of a mark unless it legally infringes theirs. The main issues to consider are whether the marks are indeed similar and confusing, whether the products are similar, whether the parties are dealing in the same channels of commerce, and whether there will be actual confusion. Beginning in 1996, there was a new factor to consider, the Federal Trademark Dilution Act which was discussed in chapter 1.

If there are enough dissimilarities, you may want to hold out and fight the opposition. If you are in your legal rights to register the mark, the person or company that filed the opposition may not want to waste the money involved in a lengthy opposition. As explained below, there are a few things you can do to see if they are serious.

Before making a decision to fight an opposition, you should consult an attorney who specializes in trademark law. His or her expertise can help you decide whether or not you should fight for your mark. Getting an opinion on your case should not be very expensive. However, having a specialist handle your opposition may be more than you can afford, especially if you are a small business. If you do have a good case but cannot afford to have such an expert handle your opposition, you can handle your own case at least long enough to see if they are bluffing, or you might be able to find a local attorney who can help you prepare the proper papers to keep your application alive.

ANSWER

The first step in fighting an opposition is to file an *answer*. An answer in a trademark opposition is similar to one filed in a lawsuit in Federal Court. The original must be filed with the Patent and Trademark Office and a copy sent to the attorney or party filing the opposition. ***It must be filed within thirty days or a default will be entered against you and your application cancelled.*** The answer is simple to prepare and can give you time to negotiate with the other side or to fight them. It may include specific defenses, for example, if you have been using the mark for many years with the full knowledge of the person opposing you. For more specific instructions, consult one of the treatises contained in the bibliography of this book. They should be available in most law libraries. *Trademark Registration Practice* by Hawes is excellent. It explains each type of answer that can be filed and exactly how to word the answer.

THREAT OF LAWSUIT

In addition to the opposition, you may get a letter from the attorney for the other party that threatens a federal lawsuit against you for infringement of their mark. While such a threat should not be taken lightly, it should be kept in mind that this may also be a bluff. Such a lawsuit would most likely be required to be filed in your district. If the other party is located a thousand miles away, it will be quite a burden for him. Even his attorneys will probably not be excited about it because they may have to hire an attorney in your area to prosecute the case.

The best way to respond to such a threat is to send a letter (preferably from your attorney) saying that you have received their letter, that you do not wish to violate their rights, that you are researching the law on the matter, and that upon determining your legal status you will immediately do whatever is necessary to comply with the law. This will show your good faith and give you the opportunity to fight back if they are legally wrong and just bluffing.

INTERROGATORIES

After your answer has been filed (or together with your answer), you can file *interrogatories* on the other side. These are written questions that the opposing party must answer under oath. These will help you get the information you need to decide if they have a good case.

For example, suppose you filed for a trademark of the word "Maple" for a computer software accounting program (in class 9), and suppose a large company that has registered Maple on machine tools (in class 7) files an opposition saying that they use software in their products. To find out if they have a good case, you should send them interrogatories similar to the following:

1. In which of your products do you use computer programs?

2. On what date did you begin using computer programs in each of your products?

3. What function does the computer program serve in each product?

4. Do you sell any of your computer programs separate and apart from the products?

5. During each of the last three years what percentage of your sales were to accountants?

The purpose of your questions should be to generate facts that the opposer's products are in a different field of commerce and that there will be no confusion between your products. In the above example, if one product will be a computer program sold to accountants and the other just the internal part of a tool used by an auto maker, there is no legal basis for denying your registration.

Interrogatories are a good way to find out how serious a party is in fighting your registration. If your questions require a lot of time to research years of records, they may decide it is not worth the trouble. If they do not take the time to answer your questions or take further action on their opposition, it may be considered abandoned. If they do go through a lot of trouble to provide the answers, they are probably serious about the opposition and you can decide if you want to come to an agreement with them or to give up the fight.

Issuance of Mark

If no opposition has been filed, if it has been filed and then abandoned or if it has been filed and the applicant prevails, then the trademark is issued. A copy of a trademark is shown on pages 99 and 100.

Sample Page from the Official Gazette

TM 68   OFFICIAL GAZETTE   APRIL 29, 1986

CLASS 16—(Continued).

SN 546.693 THOMAS INDUSTRIES INC., LOUISVILLE, KY FILED 7-8-1985.

PRO-PERFECT

FOR PAINT BRUSHES AND PAINT ROLLERS FOR USE IN INTERIOR AND EXTERIOR DECORATING AND REPAIR OF DWELLINGS AND OTHER STRUCTURES, AND OF UNITS OR PORTIONS THEREOF (U.S CL. 29).
FIRST USE 6-19-1978; IN COMMERCE 6-19-1978

SN 556,940. 13-30 CORPORATION, KNOXVILLE, TN. FILED 3-5-1985.

PARENTING ADVISER

NO CLAIM IS MADE TO THE EXCLUSIVE RIGHT TO USE "PARENTING", APART FROM THE MARK AS SHOWN
FOR POSTERS AND BOOKLETS CONTAINING ITEMS OF INTEREST TO PARENTS (U.S. CL. 38)
FIRST USE 3-14-1985; IN COMMERCE 3-14-1985.

SN 559,246. MURRAY, BILL, DBA BAM PRODUCTIONS, SHARON, PA. FILED 9-20-1985.

NO CLAIM IS MADE TO THE EXCLUSIVE RIGHT TO USE "PRODUCTIONS" APART FROM THE MARK AS SHOWN
FOR COMIC BOOKS (U.S. CL. 38).
FIRST USE 7-1-1985; IN COMMERCE 9-1-1985

SN 559,380 AUTOMOTIVE MANAGEMENT GROUP, INC., SHAWNEE MISSION, KS. FILED 9-20-1985

AUTOCOMPUTING REPORT

NO CLAIM IS MADE TO THE EXCLUSIVE RIGHT TO USE "REPORT", APART FROM THE MARK AS SHOWN.
FOR PRINTS AND PUBLICATIONS, NAMELY, A MAGAZINE PROMOTING THE USE OF MICRO AND OTHER COMPUTERS (U.S. CL. 38).
FIRST USE 8-1-1985; IN COMMERCE 8-1-1985.

CLASS 16—(Continued).

SN 560,464 HILL, MARSHA A., SILVER SPRING, MD FILED 9-27-1985

NO CLAIM IS MADE TO THE EXCLUSIVE RIGHT TO USE "CALENDARS" APART FROM THE MARK AS SHOWN.
THE LINING SHOWN IN THE MARK IS A FEATURE OF THE MARK, AND IS NOT INTENDED TO INDICATE COLOR.
FOR INDIVIDUALIZED, PERSONAL PHOTO CALENDARS (U.S CLS 37 AND 38)
FIRST USE 11-0-1984; IN COMMERCE 9-24-1985.

SN 562,719. OLMEC CORPORATION, NEW YORK, NY FILED 10-11-1985

RULERS OF THE SUN

FOR PRINTED MATTER, NAMELY, COMIC BOOKS (U.S. CL. 38).
FIRST USE 10-7-1985; IN COMMERCE 10-7-1985.

SN 563,434 SPHINX INTERNATIONAL, INC., CLEARWATER FL FILED 10-16-1985

SPHINX

FOR LAW BOOKS AND LEGAL FORMS (U.S CLS 37 AND 38)
FIRST USE 7-20-1983 IN COMMERCE 7-20-1983

SAMPLE TRADEMARK

№ 1402110

THE UNITED STATES OF AMERICA

CERTIFICATE OF REGISTRATION

This is to certify that the records of the Patent and Trademark Office show that an application was filed in said Office for registration of the Mark shown herein, a copy of said Mark and pertinent data from the Application being annexed hereto and made a part hereof,

And there having been due compliance with the requirements of the law and with the regulations prescribed by the Commissioner of Patents and Trademarks,

Upon examination, it appeared that the applicant was entitled to have said Mark registered under the Trademark Act of 1946, and the said Mark has been duly registered this day in the Patent and Trademark Office on the

PRINCIPAL REGISTER

to the registrant named herein.

This registration shall remain in force for Twenty Years unless sooner terminated as provided by law.

In Testimony Whereof I have hereunto set my hand and caused the seal of the Patent and Trademark Office to be affixed this twenty-second day of July 1986.

Donald J. Quigg

Commissioner of Patents and Trademarks

(Cover)

Int. Cl.: 16

Prior U.S. Cls.: 37 and 38

United States Patent and Trademark Office Reg. No. 1,402,110
Registered July 22, 1986

TRADEMARK
PRINCIPAL REGISTER

SPHINX

SPHINX INTERNATIONAL, INC. (FLORIDA
CORPORATION)
806 TURNER ST.
CLEARWATER, FL 33516

FOR: LAW BOOKS AND LEGAL FORMS, IN
CLASS 16 (U.S. CLS. 37 AND 38).

FIRST USE 7-20-1983; IN COMMERCE
7-20-1983.

SER. NO. 563,434, FILED 10-16-1985.

HENRY S. ZAK, EXAMINING ATTORNEY

(Inside)

PROTECTING YOUR MARK 13

Unlike patents and copyrights that have limited terms, a trademark can last forever if you continue to use it and file the proper paperwork. If you do not, it will expire.

SECTION 8 AFFIDAVIT

Your registration will be cancelled at the end of the sixth year after registration unless an affidavit that the mark is still in use has been filed with the Patent and Trademark Office during the preceding twelve month period. This is called a *Section 8 Affidavit*. (See form 10 in appendix C.) A specimen showing actual use of your mark must be included with the affidavit. In October 1999, the Trademark Law Treaty Implementation Act added a six month grace period to the twelve month period. The fee for filing the section 8 affidavit is $100.

Additional Section 8 affidavits must be filed before the end of every tenth year after registration. Since the renewal must also be filed every ten years, the Section 8 affidavit language has been combined with the renewal form.

RENEWAL

For marks registered after November 17, 1989, you must renew the mark every ten years. For marks registered prior to that date, the initial term is twenty years. Form 13 in appendix C can be used to renew a mark. As of January 2000, the renewal fee is $400.

SECTION 15
AFFIDAVIT

At any point that a mark has been in continuous use for five years, you can file a *Section 15 Affidavit*. Doing this makes the mark *incontestable* for most purposes. This means you have much stronger rights if you are ever in conflict with someone using a similar mark. (See form 11 in appendix C.) This affidavit can be filed at any time and there is no deadline. The fee is $200.

COMBINED
AFFIDAVIT

If the mark has been in continuous use for five years at the time the Section 8 affidavit is due then a combined Section 8 and Section 15 affidavit can be used. (See form 12 in appendix C.)

USE THE MARK

You must not abandon the mark. If you fail to use the mark for two years, the law says that you are assumed to have abandoned it. In such case, someone else can begin using it and eventually register it as their own. The presumption that you have abandoned it can be rebutted in special cases, such as if you have been involved in trademark litigation or if your company has been closed because of a labor strike.

GENERIC

Do not allow the mark to become *generic*. You might think that you could not be more successful than to have your name become synonymous with the product itself. This could eventually cause the loss of a trademark. "Aspirin" and "cellophane" were once trademarks owned by manufacturers of the products. Now anyone can use these words to describe their products. It is possible that Kleenex, Xerox, and Band-Aid will also lose their trademarks someday if they do not protect them. That is why you see ads stressing that it is Kleenex brand tissue.

VIGILANCE

You must protect your mark. If others start using it, you must take action to stop the use. The first step is usually to send a cease and desist letter, demanding that they immediately stop using the mark. You can also demand an accounting of all the money they made using the mark, or use it as a threat if they do not stop using the mark.

In one case, Westinghouse sued some electronics companies for selling reconditioned circuit breakers as new with the counterfeit Westinghouse trademark labels. Westinghouse lost because it had been buying the circuit breakers from the firms for years without complaining about the violation. *In re Circuit Breaker Litig.*, 106 F.3d 894 (7th Cir. 1996).

One thing to be careful about before sending a cease and desist letter is to make sure you have superior rights to the mark. We know of one case where a company sent a letter demanding that another company stop using a similar mark, but the second company had registered their trademark one month before the first company! While the designs of the two marks weren't completely similar, it would be hard for the first company to argue this since they had already claimed in their cease and desist letter that the marks were similar.

LICENSEES You must control the use of your mark by any licensees you have. If you license someone to use your mark on goods, you must reserve or exercise control over what type and quality of goods on which the mark is used.

RESTRAINT OF TRADE You must not use the mark in any attempt to restrain trade. For example, there are anti-trust laws that stipulate that the sale of one product cannot be tied to the purchase of another. If you do so with your trademark, you may lose your rights.

DESIGNATION Whenever you use a word or symbol to identify your goods, until it is registered, you should place the letters ™ next to it for goods or ℠ for services. This lets the world know that you are claiming it as your mark.

After your mark has been registered (and not before!), you should use one of the following designations next to your mark:

> ®
>
> Registered in the U.S. Patent and Trademark Office
>
> Reg. U.S. Pat. & Tm. Off.

If you fail to use one of the above, you will lose your right to collect damages from persons who infringe your mark.

CUSTOMS REGISTRATION If you have reason to believe that someone will be importing counterfeit goods with your trademark on them, you can register your mark with the U.S. Customs Service, Intellectual Property Rights Branch. To do so, you need to send them a status copy of your registration along with five copies and a fee of $190 for each class of goods for which your mark is registered. You can use form 18 in the book.

ASSIGNMENT   If you wish to transfer your rights to the trademark to someone else, an Assignment of Trademark form (form 15 in this book in appendix C) should be filed in the Patent and Trademark Office along with a cover sheet (form 16 in this book). To find out the current filing fee for an Assignment of Trademark, call 703-308-4357.

FOREIGN APPLICANTS 14

For applicants from countries that are members of the Paris Convention or the Pan-American Convention, the process of filing in the United States is simple.

These applicants may rely on their *foreign application* or *foreign registration* and do not need specimens, nor to give the dates of use, nor the manner of affixing the mark to the goods. Applicants relying on a foreign *application* must specify their priority date in the heading of their application.

The following countries are members of the above-mentioned conventions:

Algeria

Argentina

Aruba

Australia

Austria

Bahamas

Barbados

Belgium

Benin (African Union Nations) (OAPI)

Brazil

Bulgaria

Burkina Faso (OAPI)

Burundi

Cameroon (OAPI)

Canada

Central African Republic (OAPI)

Chad (OAPI)

China (People's Republic)

Commonwealth of Independent States

Congo (OAPI)

Cook Islands

Cuba

Cyprus

Czech Republic

Denmark

Dominican Republic

Egypt

Finland

France (includes all overseas Departments
    and Territories)

Gabon (OAPI)

Germany

Ghana

Greece

Guinea

Haiti

Hong Kong

Hungary

Iceland

Indonesia

Iran

Iraq

Ireland

Isle of Man

Israel

Italy

Ivory Coast (OAPI)

Japan

Jordan

Kenya

Korea, North

Korea, South

Lebanon

Libya

Liechtenstein

Luxembourg

Malawi

Malaysia

Mali (OAPI)

Malta

Mauritania (OAPI)

Mauritius

Mexico

Monaco

Mongolia

Morocco

Netherlands (Benelux Nations)

New Zealand

Niger (OAPI)

Nigeria

Norway

Philippines

Poland

Portugal

Puerto Rico

Romania

Rwanda

San Marino

Senegal (OAPI)

Slovak Republic

South Africa

Spain

Sri Lanka (formerly Ceylon)

Sudan

Suriname

Sweden

Switzerland

Syria

Tanzania

Togo (OAPI)

Tokelau Islands

Trinidad & Tobago

Tunisia

Turkey

Uganda

United Kingdom

United States (extends to all territories & possessions including Puerto Rico)

Uruguay

Vietnam

Yugoslavia

Zaire

Zambia

Zimbabwe

Applicants from countries that are not members of the above conventions, but who are from countries that are members of the Buenos Aires Convention or countries that offer reciprocal registration rights to U.S. applicants, may apply for registration based upon their foreign registration but not based upon their foreign application. These countries include:

Buenos Aires Convention:

Bolivia

Ecuador

Reciprocal Registration:

Antigua & Barbados

Belize

Brunei

Dominica

Fiji

Gambia

Grenada

Guyana

India

Jamaica

Kiribati

Lesotho

Pakistan

Saint Kitts and Nevis

Saint Lucia

Saint Vincent and the Grenadines

Seychelles

Sierra Leone

Singapore

Taiwan (Republic of China)

Tonga

Tuvalu

Venezuela

Foreign applicants must have a *domestic representative.* This is a resident of the United States on whom notices or process can be served regarding the application. The domestic representative will also receive official correspondence unless the applicant is also represented by an attorney in the United States. Designation of a domestic representative can be accomplished with form 14 included in this book, or the same wording can be added to one of the forms or to a typed application.

In the past, the Trademark Office made a big issue over whether the person signing the application was authorized to do so. In the last edition of this book, we listed the proper parties who could sign in different countries. However, the rules have been relaxed and now the person signing will be assumed to have the power to sign. Isn't it nice to see some common sense come to an area of law!

STATE
REGISTRATION 15

In addition to federal registration of trademarks, each state has its own system of registration and protection. If you plan to offer your product or service nationwide, you should proceed immediately with federal registration using the procedures either in chapter 7 or 10 of this book. If you do not yet qualify for federal registration or if your business is small and you aren't ready to pay the federal filing fee (presently $325), you can register your mark with your state. The state registration is in most cases much cheaper and simpler than federal registration. In some states, the registration fee is as low as $5 or $10.

The main benefit of either state or federal registration is that it is legal notice that you claim some interest in the mark. For federal registration, this notice applies to the entire country; for state registration, it applies only to your state. The benefit of this legal notice is that you can keep others from beginning to use the mark in the future. If someone was already using the mark before you registered it, you cannot stop them from continuing to use it (since they began first), but you can keep them from expanding their use into other areas. People thinking of using your mark will be more likely to find it in their search, and therefore less likely to use it.

A second benefit of state registration is that in some states it offers you certain legal *presumptions* if you ever have to go to court. For example, without state registration you will have to present evidence to the court that you used the name on a certain date and for certain goods. This may be expensive and time consuming since you would have to provide witnesses and records of your use. State registration (in some states) allows you to present your certificate to the court to acquire the presumption that you did use the mark on the date stated in the registration. Now it is up to the other side to disprove you used it, rather than up to you to prove it.

A more practical benefit is that your registration is a psychological advantage. Two people fighting over their unregistered claims to the mark appear to be on equal footing, but if one is registered, and *has a certificate issued by the state*, the other person will look and feel disadvantaged.

A chart is provided on the next three pages listing the trademark statutes of each state, the fee, whether they provide legal presumptions, and the types of trademarks available.

Each state has its own form, fee schedule, and local requirements. To register a name with your state, you should order forms and information directly from your state trademark office. The addresses for these offices are on the four pages following the chart.

STATE TRADEMARK LAWS

State	Trademark Statute	Fee	Presumptions	Service Marks	Trade Names
Alabama	§ 8-12-6	$30	No	Yes	Yes
Alaska	§ 45.50.010	50	No	No	No
Arizona	§ 44-1441	15	No	Yes	No
Arkansas	§ 4-71-101	50	No	Yes	No
California	CB&PC § 14200	70	Yes	Yes	No
Colorado	§ 7-70-102	50	No	Yes	No
Connecticut	CGS 621a§ 35-11a	50	Yes	Yes	No
Delaware	6 Del. C. § 3301	25	No	Yes	No
Florida	§ 495.011	87.50	Yes	Yes	No
Georgia	§ 10-1-440	15	No	Yes	No
Hawaii	§ 482-2	50	No	Yes	Yes
Idaho	§48-501	30	No	Yes	No
Illinois	Ch. 140 ¶8-22	10	Yes	Yes	No
Indiana	§ 24-2-1-1	10	No	Yes	No
Iowa	Ch. 548	10	Yes	Yes	No
Kansas	81-111	25	No	Yes	No
Kentucky	KRS 365.570	10	No	Yes	No
Louisiana	T. 9, §§ 211-224	50	No	Yes	Yes
Maine	T. 10, §§ 1521-1532	50	No	Yes	No
Maryland	art. 41, § 3-101	50	No	Yes	No

State	Trademark Statute	Fee	Presumptions	Service Marks	Trade Names
Massachusetts	c 110B	50	Yes	Yes	No
Michigan	MSA § 18.638(21)	50	Yes	Yes	No
Minnesota	333.20	50	Yes	Yes	No
Mississippi	75-25-3	50	No	Yes	No
Missouri	417.005-.031	50	No	Yes	No
Montana	30-13-301	20	No	Yes	No
Nebraska	87-210	100	No	Yes	No
Nevada	600.340 -.456	50	Yes	Yes	No
New Hampshire	c 350-A	50	No	Yes	No
New Jersey	T.t 56 c3§§ 13.1-13.9	50	No	Yes	No
New Mexico	57-3	25	No	Yes	Yes
New York	T.t.K of Arts & Cult.	50	No	Yes	No
North Carolina	80-2 to 80-15	50	No	Yes	No
North Dakota	47-22-02	30	No	No	No
Ohio	1329.54	20	No	Yes	No
Oklahoma	78-21 et seq.	50	No	Yes	No
Oregon	647.029, 035	20	Yes	Yes	No
Pennsylvania	54*1102, 1111	52	No	Yes	No
Puerto Rico	10-191, et seq.	100	Yes	Yes	Yes
Rhode Island	6-2-1 through 15	50	Yes	Yes	No
South Carolina	39-15-10 to -750	15	No	Yes	No
South Dakota	37-6	10	Yes	Yes	No

State	Trademark Statute	Fee	Presumptions	Service Marks	Trade Names
Tennessee	47-25-502	5	No	Yes	No
Texas	Cus. & com. 16.01 et seq.	50	Yes	Yes	No
Utah	Title 70 c. 3	20	Yes	Yes	No
Vermont	9-2521-75	20	No	No	No
Virginia	59.1-81	30	No	Yes	No
Washington	19.77	50	Yes	Yes	No
West Virginia	c 47, art. 2 §§ 1.2	50	Yes	No	No
Wisconsin	132.01	15	Yes	No	No
Wyoming	40-1-101	100	No	Yes	Yes

STATE TRADEMARK OFFICES

Alabama
Trademark Division
Secretary of State
State Office Building, Room 528
Montgomery, AL 36130
(205) 242-5325
http://www.sos.state.al.us/business/land.htm

Alaska
Corporations Section
Dept. of Commerce & Econ. Dev.
P. O. Box D
Juneau, AK 99811
(907) 465-2530
http://www.dced.state.ak.us/bsc/corps.htm

Arizona
Trademark Division
Secretary of State
1700 W. Washington St.
Phoenix, AZ 85007
(602) 542-6187
http://www.sosaz.com/business_services/trademarksand-
tradenames.htm

Arkansas
Trademark Division
Secretary of State
State Capitol
Little Rock, AR 72201-1094
(501) 682-3481
http://sos.state.ar.us/

California
Trademark Unit
Secretary of State
1230 "J" Street
Sacramento, CA 95814
(916) 445-9872
http://www.ss.ca.gov/

Colorado
Corporations Office
Secretary of State
1560 Broadway, #200
Denver, CO 80202
(303) 894-2251
http://www.state.co.us/gov_dir/sos/manual.html

Connecticut
Trademarks Division
Secretary of State
30 Trinity Street
Hartford, CT 06106
(860) 509-6003
http://www.sots.state.ct.us

Delaware
Trademark Filings
Division of Corporations
P. O. Box 898
Dover, DE 19903
(302) 739-3073
http://www.state.de.us/corp/index.htm

Florida
Trademark Section
Division of Corporations
P. O. Box 6327
Tallahassee, FL 32301
(904) 487-6051
http://www.dos.state.fl.us/

Georgia
Secretary of State
306 W. Floyd Towers
2 MLK Drive
Atlanta, GA 30334
(404) 656-2861
http://www.sos.state.ga.us/corporations/trademarks.htm

Hawaii
Business Registration Division
Dept. of Commerce & Consumer Affairs
1010 Richards St.
Honolulu, HI 96813
(808) 586-2730
http://www.state.hi.us/dcca/

Idaho
Trademark Division
Secretary of State
Statehouse Room 203
Boise, ID 83720
(208) 334-2300
http://www.idsos.state.id.us/tmarks/tmindex.htm

Illinois
Trademark Division
Secretary of State
111 East Monroe
Springfield, IL 62756
(217) 782-7017
http://www.sos.state.il.us/depts/bus_serv/trademrk.html

Indiana
Trademark Division
Secretary of State
State House Room 155
Indianapolis, IN 46204
(317) 232-6540
http://www.state.in.us/sos/bus_service/corps/tmgreet.html

Iowa
Corporate Division
Secretary of State
Hoover Bldg.
Des Moines, IA 50319
(515) 281-5204
http://www.sos.state.ia.us/new/sos/services.html

Kansas
Trademark Division
Secretary of State
Statehouse Bldg., Room 235N
Topeka, KS 66612
(913) 296-2034
http://www.kssos.org

Kentucky
Trademark Division
Secretary of State
Frankfort, KY 40601
(502) 564-2848
http://www.sos.state.ky.us/Trademar.htm

Louisiana
Corporation Division
Secretary of State
P. O. Box 94125
Baton Rouge, LA 70804-9125
(504) 925-4704
http://www.sec.state.la.us/

Maine
Division of Public Administration
Department of State
State House Station 101
Augusta, ME 04333
(207) 289-4195
http://www.state.me.us/sos/sos.htm

Maryland
Trademark Division
Secretary of State
State House
Annapolis, MD 21404
(301) 974-5521
http://www.dat.state.md.us/sdatweb/charter.html

Massachusetts
Trademark Division
Secretary of State
One Ashburton Place #1711
Boston, MA 02108
(617) 727-8329
http://www.state.ma.us/sec/cor/

Michigan
Corporation Division
Department of Commerce
P. O. Box 30054
Lansing, MI 48909
(517) 334-6302
http://www.cis.state.mi.us/

Minnesota
Corporation Division
Secretary of State
180 State Office Bldg.
St. Paul, MN 55155
(612) 296-3266
http://www.sos.state.mn.us

Mississippi
Trademark Division
Secretary of State
P. O. Box 1350
Jackson, MS 32915
(601) 359-1350
http://www.sos.state.ms.us/

Missouri
Trademark Division
Secretary of State
P. O. Box 778
Jefferson City, MO 65101
(314) 751-4756
http://mosl.sos.state.mo.us/bus-ser/soscor.html

Montana
Trademark Division
Secretary of State
State Capitol
Helena, MT 59620
(406) 444-3665
http://www.state.mt.us/sos/biz.htm

Nebraska
Trademark Division
Secretary of State
State Capitol Bldg.
Lincoln, NE 68509
(402) 471-4079
http://www.nol.org/home/SOS/htm/services.htm

Nevada
Trademark Division
Secretary of State
Capitol Complex
Carson City, NV 89710
(702) 687-5203
http://sos.state.nv.us/

New Hampshire
Corporation Division
Secretary of State
State House Annex
Concord, NH 03301
(603) 271-3244
http://www.state.nh.us/sos/

New Jersey
Secretary of State
State House
West State St. CN-300
Trenton, NJ 08625
(609) 984-1900
http://www.state.nj.us/state/

New Mexico
Trademark Division
Secretary of State
Capitol Bldg. Rm. 400
Santa Fe, NM 87503
(505) 827-3600
http://www.sos.state.nm.us/trade.htm

New York
Miscellaneous Records
Secretary of State
162 Washington Ave.
Albany, NY 12231
(518) 473-2492
http://www.dos.state.ny.us/corp/corpwww.html

North Carolina
Trademark Division
Secretary of State
300 N. Salisbury St.
Raleigh, NC 27611
(919) 733-4161
http://www.secstate.state.nc.us/trademrk/trade.htm

North Dakota
Trademark Division
Secretary of State
State Capitol
Bismark, ND 58505
(701) 328-4284
http://www.state.nd.us/sec/Business/businessinforegmnu.htm

Ohio
Corporations Department
Secretary of State
30 E. Broad St., 14th Fl.
Columbus, OH 43215
(614) 466-3910
http://www.state.oh.us/sos/forming.html

Oklahoma
Trademark Division
Secretary of State
101 State Capitol Bldg.
Oklahoma City, OK 73105
(405) 521-3911
http://www.state.ok.us/~sos/

Oregon
Director, Corporation Division
Secretary of State
158 - 12th Street NE
Salem, OR 97310-0210
(503) 986-2200
http://www.sos.state.or.us/corporation/corphp.htm

Pennsylvania
Corporation Bureau
Secretary of State
308 North Office Bldg.
Harrisburg, PA 17120
(717) 787-1057
http://www.dos.state.pa.us/

Puerto Rico
Trademark Division
Secretary of State
P. O. Box 3271
San Juan, PR 00904
(809) 722-2121 Ext. 337
[not yet operational]

Rhode Island
Trademark Division
Secretary of State
100 No. Main St
Providence, RI 02903
(401) 277-2340
http://www.state.ri.us/corporations/

South Carolina
Trademark Division
Secretary of State
P. O. Box 11350
Columbia, SC 29211
(803) 734-2158
http://www.leginfo.state.sc.us/secretary.html

South Dakota
Secretary of State
State Capitol Bldg.
500 East Capitol
Pierre, SD 57501
(605) 773-3537
http://www.state.sd.us/state/executive/sos/Trademarks/
Trademarks%20Intro%20Page.htm

Tennessee
Trademark Division
Secretary of State
James K. Polk Bldg. #500
Nashville, TN 37219
(615) 741-0531
http://www.state.tn.us/sos/service.htm#trademarks

Texas
Trademark Division
Secretary of State
Box 13697, Capitol Sta.
Austin, TX 78711-3697
(512) 463-5576
http://www.sos.state.tx.us/

Utah
Div. of Corporations
Heber M. Wells Bldg.
160 E. 300 South St.
Salt Lake City, UT 84111
(801) 530-4849
http://www.commerce.state.ut.us/

Vermont
Corporations Division
Secretary of State
State Office Building
Montpelier, VT 05602
(802) 828-2386
http://www.sec.state.vt.us/

Virginia
Div. of Securities & Retail Franchises
State Corp. Commission
1220 Bank Street
Richmond, VA 84111
(804) 271-9051
http://www.state.va.us/scc/index.html

Washington
Corporations Division
Secretary of State
505 E. Union St., 2nd Fl.
Olympia, WA 98504
(206) 753-7120
http://www.secstate.wa.gov/

West Virginia
Corporations Division
Secretary of State
State Capitol
Charleston, WV 25305
(304) 558-8000
http://www.state.wv.us/sos/

Wisconsin
Trademark Division
Secretary of State
P. O. Box 7848
Madison, WI 53707
(608) 266-5653
http://badger.state.wi.us/agencies/sos/trade.htm

Wyoming
Corporation Division
Secretary of State
Capitol Bldg.
Cheyenne, WY 82002
(307) 777-7311
http://soswy.state.wy.us/

OTHER PROTECTIONS 16

Although registering a mark is the best way to protect it, an unregistered mark is not without protection. If you are using a mark to distinguish your goods or services, the law protects you from others who would copy them or pass off their goods as yours. There are both federal and state laws that offer protection for marks that are unregistered:

The Lanham Act, which is the federal trademark statute, provides protection for unregistered products from *unfair trade practices*. This is contained in section 43(a) of the act. The necessary elements of a violation of the law are as follows:

☛ One party created his product through extensive time, labor, skill, or money.

☛ A second party used some aspects of the first party's product in competition with him, gaining an advantage because it was done without the cost.

☛ The second party caused commercial damage to the first.

Where these three factors are true, the first party could win damages against the second in a federal lawsuit. (Also, in some state courts, as explained below.)

However, it should be pointed out that there are many different federal courts in this country, and the rulings are not always consistent. What is illegal in one district may not be illegal in another. Sometimes two federal courts in the same state rule differently on the same issue. This may be because the lawyers didn't present the issues properly, because the judges didn't understand this obscure area of the law, or because they just interpreted the law differently.

In some ways, this protection is even stronger than trademark protection since the latter only covers you in one class of goods. Under unfair trade law, if a business has built up a reputation with its name or other aspect of its product, then it can stop anyone from *diluting* the value of its name or reputation.

As explained in chapter 1, the Federal Trademark Dilution Act greatly expanded the rights of owners of famous trademarks and made it much easier to stop similar marks.

STATE
PROTECTION
Federal trademark protection only covers goods or services that travel across state lines or to a foreign country. In some states, there are statutes similar to the Lanham Act that give state remedies for similar activities which occur within a state. In addition, states without specific statutes have court decisions, or common law, which protects businesses from unfair competition.

Some states have either dilution laws, unfair competition laws or both. If you think someone is violating your rights or you are afraid of violating someone else's, you should research your own state's laws further. On the following pages is a list of which states protect marks against dilution and unfair trade practices.

	Dilution Law	Unfair Competition Law
Alabama	§ 8-12-17	No
Alaska	§ 45.50.180	No
Arizona	No	No
Arkansas	§ 4-71-113	No
California	B&PC § 14300	B&PC § 17200
Colorado	No	No
Connecticut	621a, § 35-11i	No
Delaware	6 Del.C. § 3313	6 Del.C. § 2531
District of Columbia	No	No
Florida	§ 495.151	§ 501.201 et seq
Georgia	10-1-451(b)	23-2-55
Hawaii	No	480-2; 487-5(5)
Idaho	48-512	No
Illinois	140, ¶22	No
Indiana	No	No
Iowa	548.11(2)	No
Kansas	No	No
Kentucky	No	No
Louisiana	51:223.1	51:1401 et seq
Maine	10 MRSA §1530	No
Maryland	No	§ 13-301
Massachusetts	CH. 110B, §12	No
Michigan	No	No
Minnesota	§ 325D	No
Mississippi	§ 75-25-25	No
Missouri	§ 417.061(1)	No
Montana	§ 30-13-334	No

	Dilution Law	Unfair Competition Law
Nebraska	§ 59-1601	No
Nevada	No	No
New Hampshire	§ 350-A:12	§ 358-A:2
New Jersey	13PL1995 c.171	§ 56:4-1
New Mexico	57-3-10	No
New York	G.B.L. § 368-d	No
North Carolina	No	No
North Dakota	No	No
Ohio	No	No
Oklahoma	No	No
Oregon	§ 647.107	§ 646.608(1)(b)
Pennsylvania	54PCSA § 1124	No
Rhode Island	§ 6-2-12	No
South Carolina	§ 39-15-1105(2)	No
South Dakota	No	§ 37-6-2
Tennessee	§ 47-25-512	No
Texas	B&CC 16.29	PC § 32.42
Utah	No	No
Vermont	No	9 VSA § 2532
Virginia	No	§ 18.2-499
Washington	RCW 19.77	No
West Virginia	§ 47-2-13	§ 47-11A-1 et seq
Wisconsin	No	No
Wyoming	No	No

CHECKLISTS 17

The checklists on the following pages will help you to keep track of the steps in registering your trademark. To visualize the process, you can also refer to the flow chart in the introduction.

PRESENT USE CHECKLIST

Choose the mark—
    ☐ Does it project a positive image?
    ☐ Will it have continued value?
    ☐ Will it be useful in other states and countries?
    ☐ Make sure it is not merely descriptive
    ☐ Make sure it is not forbidden

Search the mark—
    ☐ Preliminary search of locally available sources
    ☐ Search Patent and Trademark Office
    ☐ Common Law search (phone books, directories, etc.)

Use the mark—
    ☐ Use it *in commerce*
    ☐ Use must be bona fide

Prepare the application—
    ☐ Choose right form–trademark/collective/membership/certification
    ☐ Create drawing
    ☐ Obtain specimens of mark as actually used

Send it all in—
    ☐ Application–signed
    ☐ Drawing
    ☐ Specimen
    ☐ Fee
    ☐ Self-addressed stamped envelope

Answer correspondence—
    ☐ Follow all instructions to the letter
    ☐ If opposition–fight it or abandon it

After mark is issued—
    ☐ File § 8 affidavit after fifth anniversary and before sixth anniversary
    ☐ If use has been continuous, file § 8 & § 15 affidavit instead
    ☐ Use ® symbol with your mark
    ☐ Do not abandon your mark
    ☐ Do not allow it to become generic
    ☐ Do not allow others to infringe your mark
    ☐ Control the use of your mark by licensees
    ☐ Do not use your mark in restraint of trade
    ☐ File § 8 affidavit every ten years
    ☐ Renew mark every ten years

Intent to Use Checklist

Choose the mark—
- ☐ Does it project a positive image?
- ☐ Will it have continued value?
- ☐ Will it be useful in other states and countries?
- ☐ Make sure it is not merely descriptive
- ☐ Make sure it is not forbidden

Search the mark—
- ☐ Preliminary search of locally available sources
- ☐ Search Patent and Trademark Office
- ☐ Common Law search (phone books, directories, etc.)

Prepare the application—
- ☐ Choose right form–trademark/collective/membership/certification
- ☐ Create drawing

Send it in—
- ☐ Application–signed
- ☐ Drawing
- ☐ Fee
- ☐ Self-addressed stamped envelope

Use the mark—
- ☐ Use it *in commerce*
- ☐ Use must be bona fide

Submit proof of use—
- ☐ Amendment to Allege Use, or Statement of Use
- ☐ Include specimen of use
- ☐ Fee

Answer correspondence—
- ☐ Follow all instructions to the letter
- ☐ If opposition–fight it or abandon it

After mark is issued—
- ☐ File § 8 affidavit after fifth anniversary and before sixth anniversary
- ☐ If use has been continuous, file § 8 & § 15 affidavit instead
- ☐ Use ® with your mark
- ☐ Do not abandon your mark
- ☐ Do not allow it to become generic
- ☐ Do not allow others to infringe your mark
- ☐ Control the use of your mark by licensees
- ☐ Do not use your mark in restraint of trade
- ☐ File § 8 affidavit every ten years
- ☐ Renew mark every ten years

Appendix A
Trademark Classifications

Before searching your mark, you will need to determine into which classification the goods or services will fall. On the following pages are the International Classifications.

If your goods or services clearly fall into one classification, you can write it on your application. For a much longer listing, you can go to the USPTO's website at: http://www.uspto.gov/web/offices/tac/doc/gsmanual/.

However, if you are not sure which class is appropriate for your goods or services, you can leave it blank and the Patent and Trademark Office will fill it in for you.

INTERNATIONAL CLASSIFICATIONS

—Goods—

1. Chemicals used in industry, science, photography, as well as in agriculture, horticulture, and forestry; unprocessed artificial resins; unprocessed plastics; manures; fire extinguishing compositions; tempering and soldering preparations; chemical substances for preserving foodstuffs; tanning substances; adhesives used in industry.

2. Paints, varnishes, lacquers; preservatives against rust and against deterioration of wood; colourants; mordants; raw natural resins; metals in foil and powder form for painters, decorators, printers and artists.

3. Bleaching preparations and other substances for laundry use; cleaning, polishing, scouring and abrasive preparations; soaps; perfumery, essential oils, cosmetics, hair lotions; dentifrices.

4. Industrial oils and greases; lubricants; dust absorbing, wetting and binding compositions; fuel (including motor spirit) and illuminants; candles, wicks.

5. Pharmaceutical, veterinary, and sanitary preparations; dietetic substances adapted for medical use, food for babies; plasters, materials for dressings, material for stopping teeth, dental wax, disinfectants; preparations for destroying vermin; fungicides, herbicides.

6. Common metals and their alloys; metal building materials; transportable buildings of metal; materials of metal for railway racks; non-electric cables and wires of common metal; iron-mongery, small items of metal hardware; pipes and tubes of metal; safes; goods of common metal not included in other classes; ores.

7. Machines and machine tools; motors (except for land vehicles); machine coupling and belting (except for land vehicles); agricultural implements; incubators for eggs.

8. Hand tools and implements (hand-operated); cutlery; side arms; razors.

9. Scientific, nautical, surveying, electric, photographic, cinematographic, optical, weighing, measuring, signalling, checking (supervision), life-saving and teaching apparatus and instruments; apparatus for recording transmission or reproduction of sound or images; magnetic data carriers, recording discs; automatic vending machines and mechanisms for coin-operated apparatus; cash registers, calculating machines, data processing equipment and computers; fire-extinguishing apparatus.

10. Surgical, medical, dental, and veterinary apparatus and instruments, artificial limbs, eyes and teeth; orthopedic articles; suture materials.

11. Apparatus for lighting, heating, steam generating, cooking, refrigerating, drying, ventilating, water supply, and sanitary purposes.

12. Vehicles; apparatus for locomotion by land, air or water.

13. Firearms; ammunition and projectiles; explosives; fireworks.

14. Precious metals and their alloys and goods in precious metals or coated therewith, not included in other classes; jewelry, precious stones; horological and other chronometric instruments.

15. Musical instruments.

16. Paper and cardboard and goods made from these materials, not included in other classes; printed matter; bookbinding material; photographs; stationery; adhesives for stationery or household purposes; artists' materials; paint brushes; typewriters and office requisites (except furniture); instructional and teaching material (except apparatus); plastic materials for packaging (not included on other classes); playing cards; printers' type; printing blocks.

17. Rubber, gutta-percha, gum, asbestos, mica and goods made from these materials and not included in other classes; plastics in extruded form for use in manufacture; packing, stopping and insulating materials; flexible pipes, not of metal.

18. Leather and imitations of leather, and goods made from these materials and not included in other classes; animal skins, hides; trunks and travelling bags; umbrellas, parasols and walking sticks; whips, harness and saddlery.

19. Building materials (non-metallic); non-metallic rigid pipes for building; asphalt, pitch and bitumen; non-metallic transportable buildings; monuments, not of metal.

20. Furniture, mirrors, picture frames, goods (not included in other classes) of wood, cork, reed cane, wicker, horn, bone, ivory, whalebone, shell, amber, mother-of-pearl, meerschaum and substitutes for all these materials, or of plastics.

21. Household or kitchen utensils and containers (not of precious metal or coated therewith); combs and sponges; brushes (except paint brushes); brush-making materials; articles for cleaning purposes; steel wool; unworked or semi-worked glass (except glass used in building); glassware, porcelain and earthenware, not included in other classes.

22. Ropes, string, nets, tents, awnings, tarpaulins, sails, sacks; and bags (not included in other classes); padding and stuffing materials (except of rubber or plastics); raw fibrous textile materials.

23. Yarns and threads, for textile use.

24. Textile and textile goods, not included in other classes; bed and table covers.

25. Clothing, footwear, headgear.

26. Lace and embroidery, ribbons and braid; buttons, hooks and eyes, pins and needles; artificial flowers.

27. Carpets, rugs, mats and matting; linoleum and other materials for covering existing floors; wall hangings (non-textile).

28. Games and playthings; gymnastic and sporting articles not included in other classes; decorations for Christmas trees.

29. Meats, fish, poultry and game; meat extracts; preserved, dried and cooked fruits and vegetables; jellies, jams; eggs, milk and milk products; edible oils and fats; salad dressings; preserves.

30. Coffee, tea, cocoa, sugar, rice, tapioca, sago, artificial coffee; flour, and preparations made from cereals, bread, pastry and confectionery, ices; honey, treacle; yeast, baking-powder; salt, mustard, vinegar, sauces, (except salad dressings) spices; ice.

31. Agricultural, horticultural and forestry products and grains not included in other classes; living animals; fresh fruits and vegetables; seeds, natural pants and flowers; foodstuffs for animals, malt.

32. Beers; mineral and aerated waters and other non-alcoholic drinks; fruit drinks and fruit juices; syrups and other preparations for making beverages.

33. Alcoholic beverages (except beers).

34. Tobacco; smokers' articles; matches.

—Services—

35. Advertising; business management; business administration; office functions.

36. Insurance; financial affairs; monetary affairs; real estate affairs.

37. Building construction; repair; installation services.

38. Telecommunications.

39. Transport; packaging and storage of goods; travel arrangement.

40. Treatment of materials.

41. Education; providing of training; entertainment; sporting and cultural activities.

42. Providing of food and drink; temporary accommodation; medical, hygienic and beauty care; veterinary and agricultural services; legal services; scientific and industrial research; computer programming, services that cannot be classified in other classes.

APPENDIX B
SAMPLE FILLED-IN FORMS

The following pages contain the forms mentioned throughout this book. They may be either removed from this book and used, or photocopied.

<u>ADOLPH BUSH</u>
[Your Name]
<u>123 MALT LANE</u>
[Address]
<u>EAST ST. LOUIS, IL 61201</u>
[City, State Zip]
<u>217-555-1212</u>
[Phone]

January 5, 2000
[Date]

Hon Commissioner of Patents and Trademarks
Washington, DC 20231

Re: In Use Trademark Application

Sir:

Enclosed for filing please find trademark application papers of <u>BUSH BROTHERS</u> <u>_____</u> for the mark "<u>_____OLD NAG_____</u>" in International Class <u>_32_</u> together with the drawing, a specimen and the filing fee of $325.

A self-addressed, stamped envelope is enclosed for return of the receipt.

Respectfully submitted,

Adolph Bush
<u>_____</u>
[Your Name]

AMALGAMATED DANCERS' UNION, LOCAL 55, INC.
[Your Name]

257 83RD ST.
[Address]

NEW YORK, NY 10011
[City, State Zip]

(212)555-0000
[Phone]

February 10, 1999
[Date]

Hon Commissioner of Patents and Trademarks
Washington, DC 20231

Re: Intent-to-Use Trademark Application

Sir:

Enclosed for filing please find trademark application papers of AMALGAMATED
 UNION, LOCAL 55, INC. for the mark " ADU - LOCAL 55 " in International
Class 200 together with the drawing and the filing fee of $325.

A self-addressed, stamped envelope is enclosed for return of the receipt.

Respectfully submitted,

_____*John Doe*_____
[Your Name]

~TRADEMARK/SERVICE MARK APPLICATION (15 U.S.C. §§ 1051, 1126(d)&(e))~

~To the Assistant Commissioner for Trademarks~

<APPLICANT INFORMATION>

<Name>   BUSH BROTHERS

<Street>   123 MALT LANE

<City>   EAST ST. LOUIS

<State>   ILLINOIS

<Country> USA

<Zip/Postal Code>   61201

<Telephone Number>   206-123-4567

<Fax Number>   206-123-4568

<e-mail Address>   BUSH@BUSHBRO.COM

<APPLICANT ENTITY INFORMATION>~*Select only ONE*~

<Individual: Country of Citizenship>

<Corporation: State/Country of Incorporation>

<Partnership: State/Country under which Organized>   ILLINOIS

<Name(s) of General Partner(s) & Citizenship/Incorporation>

<Other Entity Type: Specific Nature of Entity>

<State/Country under which Organized>

<TRADEMARK/SERVICE MARK INFORMATION>

<Mark>   OLD NAG

<Typed Form>~*Enter YES, if appropriate*~ YES

~***DISPLAY THE MARK** that you want to register on a separate piece of paper (even if simply a word(s)). Please see additional **HELP** instructions.*~

<BASIS FOR FILING AND GOODS/SERVICES INFORMATION>

<**Use in Commerce: Section 1(a)**>~*Applicant is using the mark in commerce on or in connection with the below-identified goods and/or services (15 U.S.C § 1051(a)).*~

<International Class Number(s)> 32

<Listing of Goods and/or Services>~***List in ascending numerical class order. Please see sample in HELP instructions.***~

BEER AND MALT LIQUOR

> *ONLY FILL IN THIS SECTION IF YOU HAVE **ALREADY USED** THE MARK IN COMMERCE*

<Date of First Use Anywhere> 05/16/1966

<Date of First Use in Commerce> 07/21/2001

~*Submit one (1) **SPECIMEN** for each international class showing the mark as used in commerce.*~

PTO Form 1478 (REV 5/99)
OMB Control No. 0651-0009 (Exp. 8/31/2001)

<Intent to Use: Section 1(b)>~*Applicant has a bona fide intention to use the mark in commerce on or in connection with the below-identified goods and/or services (15 U.S.C. § 1051(b)).*~

<International Class Number(s)> 32

<Listing of Goods and/or Services>~*List in ascending numerical class order. Please see sample in HELP instructions.*~

```
32:  BEER AND MALT LIQUOR
```

ONLY FILL IN THIS
SECTION IF YOU HAVE
NOT USED THE MARK
IN COMMERCE

<Foreign Priority: Section 44(d)>~*Applicant has a bona fide intention to use the mark in commerce on or in connection with the below-identified goods and/or services, and asserts a claim of priority based upon a foreign application in accordance with 15 U.S.C. § 1126(d).*~

<International Class Number(s)> 32

<Listing of Goods and/or Services>~*List in ascending numerical class order. Please see sample in HELP instructions.*~

```
32:  BEER AND MALT LIQUOR
```

ONLY FILL IN THIS
SECTION IF YOU HAVE
FILED AN APPLICATION
FOR THIS TRADEMARK IN
ANOTHER COUNTRY

<Country of Foreign Filing>   GERMANY

<Foreign Application Number> TR-1234567890

<Date of Foreign Filing>   02/18/1999

<Foreign Registration: Section 44(e)>~*Applicant has a bona fide intention to use the mark in commerce on or in connection with the below-identified goods and/or services based on registration of the mark in applicant's country of origin.*~

<International Class Number(s)> 32

<Listing of Goods and/or Services>~*List in ascending numerical class order. Please see sample in HELP instructions.*~

```
32:  BEER AND MALT LIQUOR
```

ONLY FILL IN THIS
SECTION IF YOU HAVE
REGISTERED THIS
TRADEMARK IN
ANOTHER COUNTRY

<Country of Foreign Registration>   GERMANY

<Foreign Registration Number>   TR-1234567890

<Foreign Registration Date>   09/29/1999

<Foreign Registration Renewal Date>   09/29/2009

<Foreign Registration Expiration Date> 09/29/2019

~*Submit foreign registration certificate or a certified copy of the foreign registration, in accordance with 15 U.S.C. §1126(e).*~

<FEE INFORMATION>

$325.00 x <Number of Classes>   1   = <Total Filing Fee Paid> $325.00

< SIGNATURE INFORMATION>

~Applicant requests registration of the above-identified mark in the United States Patent and Trademark Office on the Principal Register established by Act of July 5, 1946 (15 U.S.C. § 1051 et seq.) for the above-identified goods and/or services.

The undersigned, being hereby warned that willful false statements and the like so made are punishable by fine or imprisonment, or both, under 18 U.S.C. § 1001, and that such willful false statements may jeopardize the validity of the application or any resulting registration, declares that he/she is properly authorized to execute this application on behalf of the applicant; he/she believes the applicant to be the owner of the trademark/service mark sought to be registered, or, if the application is being filed under 15 U.S.C. § 1051(b), he/she believes applicant to be entitled to use such mark in commerce; to the best of his/her knowledge and belief no other person, firm, corporation, or association has the right to use the mark in commerce, either in the identical form thereof or in such near resemblance thereto as to be likely, when used on or in connection with the goods/services of such other person, to cause confusion, or to cause mistake, or to deceive; and that all statements made of his/her own knowledge are true; and that all statements made on information and belief are believed to be true.~

~Signature~ *Joseph Bush*

<Date> 02/02/2002

<Name>JOSEPH BUSH

<Title> PARTNER

<CONTACT INFORMATION>

<Name>   JOSEPH BUSH

<Company/Firm Name>   BUSH BROTHERS

<Street>   123 MALT LANE

<City>   EAST ST. LOUIS

<State>   ILLINOIS

<Country> USA

<Zip/Postal Code>   61201

<Telephone Number>   206-123-4567

<Fax Number>   206-123-4568

<e-Mail Address>   JBUSH@BUSHBRO.COM

The information collected on this form allows the PTO to determine whether a mark may be registered on the Principal or Supplemental Register, and provides notice of an applicant's claim of ownership of the mark or bona fide intent to use the mark in commerce. Responses to the request for information are required to obtain the benefit of a registration on the Principal or Supplemental Register. 15 U.S.C. §§1051 et seq. and 37 C.F.R. Part 2. All information collected will be made public. Gathering and providing the information will require an estimated seventeen to twenty-three minutes. Please direct comments on the time needed to complete this form, and/or suggestions for reducing this burden to the Chief Information Officer, U.S. Patent and Trademark Office, U.S. Department of Commerce, Washington D.C. 20231. Please note that the PTO may not conduct or sponsor a collection of information using a form that does not display a valid OMB control number. (See bottom left side of this form).

COLLECTIVE TRADEMARK/SERVICE MARK APPLICATION, PRINCIPAL REGISTER, WITH DECLARATION	MARK (Word(s) and/or Design) U.S.U.S.	CLASS NO. (If known) 18

TO THE ASSISTANT COMMISSIONER FOR TRADEMARKS:

APPLICANT'S NAME:	UNITED STATES UMBRELLA SOCIETY, INC.
APPLICANT'S MAILING ADDRESS: (Display address exactly as it should appear on registration)	101 WHARF STREET SEATTLE, WA 97654

APPLICANT'S ENTITY TYPE: (Check one and supply requested information)

	Individual - Citizen of (Country):
	Partnership - State where organized (Country, if appropriate): _____ Names and Citizenship (Country) of General Partners: _____
X	Corporation - State (Country, if appropriate) of Incorporation: WASHINGTON
	Other (Specific Nature of Entity and Domicile):

GOODS AND/OR SERVICES:

Applicant requests registration of the collective mark shown in the accompanying drawing in the United States Patent and Trademark Office on the Principal Register established by the Act of July 5, 1946 (15 U.S.C. 1051 et. seq., as amended) for the following goods/services (**SPECIFIC GOODS AND/OR SERVICES MUST BE INSERTED HERE**) _____

UMBRELLAS

BASIS FOR APPLICATION: (Check boxes which apply, but never both the first AND second boxes, and supply requested information related to each box checked.)

[ ]	Applicant is exercising legitimate control over the use of the mark in commerce by its members on or in connection with the above identified goods/services. (15 U.S.C. 1051 (a) and 1054, as amended.) Three specimens showing the mark as used by the members in commerce are submitted with this application. • Date of first use of the mark in commerce which the U.S. Congress may regulate (for example, interstate or between the U.S. and a foreign country): _____ • Specify the type of commerce: _____ (for example, interstate or between the U.S. and a specified foreign country) • Date of first use anywhere (the same as or before use in commerce date): _____ • Specify intended manner or mode of use of mark on or in connection with the goods/services:_____ (for example, trademark is applied to labels, service mark is used in advertisements)
[ ]	Applicant has a bona fide intention to exercise legitimate control over the use of the mark in commerce by its members on or in connection with the above-identified good/services. (15 U.S. C. 1051(b) and 1054, as amended.) • Specify intended manner or mode of use of mark on or in connection with the goods/services: _____ (for example, trademark will be applied to labels, service mark will be used in advertisements)
[X]	Applicant has a bona fide intention to exercise legitimate control over the use of the mark in commerce by its members on or in connection with the above identified goods/services, and asserts a claim of priority based upon a foreign application in accordance with 15 U.S.C. 1126(d), as amended. • Country of foreign filing: GREAT BRITAIN • Date of foreign filing: AUGUST 9, 1999
[ ]	Applicant has a bona fide intention to exercise legitimate control over the use of the mark in commerce by its members on or in connection with the above identified goods/services and, accompanying this application, submits a certification or certified copy of a foreign registration in accordance with 15 U.S.C 1126(e), as amended. • Country of registration: _____ • Registration number: _____

NOTE: Declaration, on Reverse Side, MUST be Signed

PTO Form 1478(a) (REV 6/96)   U.S. DEPARTMENT OF COMMERCE/Patent and Trademark Office
OMB No. 0651-0009 (Exp. 06/30/98) There is no requirement to respond to this collection of information unless a currently valid OMB Number is displayed.

COLLECTIVE MEMBERSHIP MARK APPLICATION, PRINCIPAL REGISTER, WITH DECLARATION	MARK (Word(s) and/or Design) ADU – LOCAL 55	CLASS NO. 200

TO THE ASSISTANT COMMISSIONER FOR TRADEMARKS:

APPLICANT'S NAME: AMALGAMATED DANCERS' UNION, LOCAL 55, INC.

APPLICANT'S MAILING ADDRESS: 257 83RD ST.
NEW YORK, NY 10011

(Display address exactly as it should appear on registration)

APPLICANT'S ENTITY TYPE: (Check one and supply requested information)

	Individual - Citizen of (Country):
	Partnership - State where organized (Country, if appropriate): _____ Names and Citizenship (Country) of General Partners:
X	Corporation - State (Country, if appropriate) of Incorporation: NEW YORK
	Other (Specify Nature of Entity and Domicile):

Applicant requests registration of the collective membership mark shown in the accompanying drawing in the United States Patent and Trademark Office on the Principal Register established by the Act of July 5, 1946 (15 U.S.C. 1051 et. seq., as amended) to indicate membership in a(n): _____
   AMALGAMATED DANCERS' UNION, LOCAL 55, WHICH IS A LABOR UNION

(Specify the type or nature of the organization, for example, a social club, labor union, political society, or an association of real estate brokers.)

BASIS FOR APPLICATION: (Check boxes which apply, but **never both the first AND second boxes**, and supply requested information related to each box checked.)

[ ]	Applicant is exercising legitimate control over the use of the mark in commerce by its members to indicate membership. (15 U.S.C. 1051(a) and 1054 as amended.) Three specimens showing the mark as used by the members in commerce are submitted with this application.
	• Date of first use of the mark by members in commerce which the U.S. Congress may regulate (for example, interstate or between the U.S. and a specified foreign country): _____
	• Specify the type of commerce: _____ (for example, interstate or between the U.S. and a specified foreign country)
	• Date of first use anywhere (the same as or before use in commerce date): _____
	• Specify manner or method of using mark to indicate membership: _____
	(for example, mark is applied to membership cards, certificates, window decals)

[X]	Applicant has a bona fide intention to exercise legitimate control over the use of the mark in commerce by its members to indicate membership. (15 U.S.C. 1051(b) and 1054, as amended.)
	• Specify intended manner or method of using mark to indicate membership _____ MARK WILL BE EMBROIDERED PATCHES WORN ON DANCERS' TUTUS
	(for example, mark is applied to membership cards, certificates, window decals)

[ ]	Applicant has a bona fide intention to exercise legitimate control over the use of the mark in commerce by its members to indicate membership, and asserts a claim of priority based upon a foreign application in accordance with 15 U.S.C. 1126(d), as amended.
	• Country of foreign filing: _____ • Date of foreign filing: _____

[ ]	Applicant has a bona fide intention to exercise legitimate control over the use of the mark in commerce by its members to indicate membership and, accompanying this application, submits a certification or certified copy of a foreign registration in accordance with 15 U.S.C 1126(e), as amended.
	• Country of registration: _____ • Registration number: _____

NOTE: Declaration, on Reverse Side, MUST be Signed

PTO Form 4.8
OMB No. 0651-0009 (Exp. 06/30/98) There is no requirement to respond to this collection of information unless a currently valid OMB Number is displayed.
U.S. DEPARTMENT OF COMMERCE/Patent and Trademark Office

CERTIFICATION MARK APPLICATION, PRINCIPAL REGISTER, WITH DECLARATION	MARK (Word(s) and/or Design) F-L
	CLASS ☒ A. Goods [ ] B. Services

TO THE ASSISTANT COMMISSIONER FOR TRADEMARKS:

APPLICANT'S NAME: FRED'S LABS, INC.

APPLICANT'S MAILING ADDRESS: 1209 EAST END RD.

(Display address exactly as it should appear on registration)   SELMA, AL 30110

APPLICANT'S ENTITY TYPE: (Check one and supply requested information)

	Individual - Citizen of (Country):
	Partnership - State where organized (Country, if appropriate): _____ Names and Citizenship (Country) of General Partners: _____
X	Corporation - State (Country, if appropriate) of Incorporation: ALABAMA
	Other (Specify Nature of Entity and Domicile):

GOODS AND/OR SERVICES:

Applicant requests registration of the certification mark shown in the accompanying drawing in the United States Patent and Trademark Office on the Principal Register established by the Act of July 5, 1946 (15 U.S.C. 1051 et. seq., as amended) for the following goods/services (**SPECIFIC GOODS AND/OR SERVICES MUST BE INSERTED HERE**): _____
    ELECTRICAL APPLIANCES

The certification mark, as used (or, if filing under 15 U.S.C. 1051(b), intended to be used) by authorized persons, certifies (or, if filing under 15 U.S.C. 1051(b), is intended to certify): IT HAS BEEN TESTED BY FRED'S LABS, INC. AND FOUND TO BE SAFE TO
(for example, a particular regional origin of the goods, a characteristic of the goods or services, that labor was performed by a particular group)

BASIS FOR APPLICATION: (Check boxes which apply, but never both the first AND second boxes, and supply requested information related to each box checked.)

☒	Applicant is exercising legitimate control over the use of the certification mark in commerce on or in connection with the above identified goods/services. (15 U.S.C. 1051(a) and 1054 as amended.) Three specimens showing the mark as used by authorized persons in commerce are submitted with this application.
	    • Date of first use anywhere of the mark by authorized person in commerce which the U.S. Congress may regulate (for example, interstate or between the U.S. and a specified foreign country): JANUARY 26, 1999
	    • Specify the type of commerce: INTERSTATE
	     (for example, interstate or between the U.S. and a specified foreign country)
	    • Date of first use anywhere by an authorized person (the same as or before use in commerce date): JANUARY 19, 1963
	    • Specify manner of using mark on or in connection with the goods/services:
	     MARK IS APPLIED TO LABELS ATTACHED TO THE GOODS
	     (for example, trademark is applied to labels, service mark is used in advertisements)

[ ]	Applicant has a bona fide intention to exercise legitimate control over the use of the certification mark in commerce on or in connection with the above identified goods/services. (15 U.S.C. 1051(b) and 1054, as amended.)
	    • Specify intended manner or method of using mark on or in connection with the goods/services: _____
	     (for example, trademark will be applied to labels, service mark will be used in advertisements)

[ ]	Applicant has a bona fide intention to exercise legitimate control over the use of the certification mark in commerce on or in connection with the above identified goods/services, and asserts a claim of priority based upon a foreign application in accordance with 15 U.S.C. 1126(d), as amended. •Country of foreign filing: _____ •Date of foreign filing: _____
[ ]	Applicant has a bona fide intention to exercise legitimate control over the use of the certification mark in commerce on or in connection with the above identified goods/services and, accompanying this application, submits a certification or certified copy of a foreign registration in accordance with 15 U.S.C 1126(e), as amended. •Country of registration: _____ • Registration number: _____

PTO Form 4.9
OMB No. 0651-0009 (Exp. 06/30/98) There is no requirement to respond to this collection of information unless a currently valid OMB Number is displayed.

U.S. DEPARTMENT OF COMMERCE/Patent and Trademark Office

140

~TRADEMARK/SERVICE MARK ALLEGATION OF USE (Statement of Use/ Amendment to Allege Use) (15 U.S.C. § 1051(c) or (d))~

~To the Assistant Commissioner for Trademarks~

<TRADEMARK/SERVICEMARK INFORMATION>

<Mark> SAHARA

<Serial Number> 1357924680

<APPLICANT INFORMATION>

<Name> MARTIN SAND

<Street> 123 DESERT PARKWAY

<City> MOAB

<State> UTAH

<Country> USA

<Zip/Postal Code> 80123

<NOTICE OF ALLOWANCE INFORMATION>

<Notice of Allowance> YES ~*Enter YES if you are filing the Allegation of Use **after** a Notice of Allowance has issued. If not, enter NO.*~

<GOODS AND/OR SERVICES INFORMATION>

<All Goods and/or Services in Application/Notice of Allowance>~*The owner is using the mark in commerce on or in connection with all goods and/or services listed in the application or Notice of Allowance. If not, list in the next section the goods and/or services not in use to be deleted.*~

<Goods and/or Services Not in Use to be **Deleted**>~*In the following space, list only those goods and/or services (and/or entire classes(es)) appearing in the application or Notice of Allowance for which the owner is **not** using the mark in commerce.* **LEAVE THIS SPACE BLANK IF THE OWNER IS USING THE MARK ON OR IN CONNECTION WITH ALL THE GOODS AND/OR SERVICES LISTED IN THE APPLICATION OR NOTICE OF ALLOWANCE.**~

<USE INFORMATION>

<Date of First Use Anywhere> 08/21/2003

<Date of First Use in Commerce> 12/21/2003

<OPTIONAL - REQUEST TO DIVIDE INFORMATION>

<Request to Divide> NO ~*Enter YES if you are you submitting a request to divide with this document. If not, enter NO.*~

PTO Form 1553 (REV 5/99)
OMB Control No. 0651-0009 (Exp. 8/31/2001)

U.S. DEPARTMENT OF COMMERCE/Patent and Trademark Office
There is no requirement to respond to this collection of information
unless a currently valid OMB number is displayed.

~REQUEST FOR EXTENSION OF TIME TO FILE A STATEMENT OF USE (15 U.S.C. § 1051(d))~

~To the Assistant Commissioner for Trademarks~

<TRADEMARK/SERVICEMARK INFORMATION>

<Mark> SPIDER

<Serial Number> 2468013579

<APPLICANT INFORMATION>

<Name> SPIDER WEBBING, INC.

<Street> 123 JOHNSON RD

<City> JACKSON

<State> MISSISSIPPI

<Country> USA

<Zip/Postal Code> 80123

<NOTICE OF ALLOWANCE INFORMATION>

<Notice of Allowance Mailing Date> 06/30/2001 ~Enter date in the format MM/DD/YYYY.~

<GOODS AND/OR SERVICES INFORMATION>

<All Goods and/or Services in Notice of Allowance>~*The applicant has a continued bona fide intention to use the mark in commerce on or in connection with all the goods and/or services listed in the Notice of Allowance. If not, list in the next section the goods and/or services to be deleted.*~

<Goods and/or Services to be **Deleted**>~*In following space, list only those goods/services (or entire classes(es)) appearing in the Notice of Allowance for which the applicant does **not** have a continued bona fide intention to use the mark in commerce.* **LEAVE THIS SPACE BLANK IF THE APPLICANT <u>DOES</u> HAVE A CONTINUED BONA FIDE INTENTION TO USE THE MARK IN COMMERCE ON OR IN CONNECTION WITH <u>ALL</u> GOODS/ SERVICES LISTED IN THE NOTICE OF ALLOWANCE.*~

<EXTENSION REQUEST INFORMATION>~

<Number of Extension Request> 3rd ~*Enter which request (1st, 2nd, 3rd, 4th or 5th) this is following the mailing of the Notice of Allowance.*~

<ONGOING EFFORTS TO USE MARK IN COMMERCE>~*Applies to 2nd, 3rd, 4th & 5th extension requests only.*~

~*The applicant has made the following ongoing efforts to use the mark in commerce on or in connection with those goods and/or services for which use of the mark in commerce has not yet been made.*~

<Explanation> THE MANUFACTURER HAD DELAYED IN DELIVERING SOME OF THE GOODS. SAID GOODS WILL BE USED IN COMMERCE AS SOON AS DELIVERED.

TRADEMARK

UNITED STATES DEPARTMENT OF COMMERCE
PATENT AND TRADEMARK OFFICE

Applicant SAM STREETCAR

Mark JAYMEE

Serial Number 19234167

Trademark Law Office D206

Trademark Attorney STAN R. KUNTZ, III

Filed 01/29/2000

Commissioner of Patents and Trademarks
Washington, D.C. 20231

REQUEST TO DIVIDE APPLICATION

The applicant hereby requests that the application identified above be divided as follows:

Please retain in the original application the following goods/services—(use the language of the original application insofar as possible)

BOOKS

Please include in the new, divided application the following goods/services—(They should be different from and should not overlap, those remaining in the original application.)

AUDIO CASSETTES

(check one)

[X] Enclosed is a check in payment of the filing fee for the divided application

[ ] The divided application includes all goods or services in a single class presented in the original, parent application; therefore the applicant submits that no filing fee is due or required.

DATED: 02/02/2002

Respectfully,

Sam Streetcar

Telephone Number: 312-555-1212
Address: 124 HERBERTSON RD., MUNDY, IL 12456

~DECLARATION OF USE OF MARK IN COMMERCE UNDER § 8 (15 U.S.C. § 1058)~

~To the Assistant Commissioner for Trademarks~

<TRADEMARK/SERVICE MARK INFORMATION>

<Mark>   SAHARA
<Registration Number> 1357924680
<Registration Date> 02/02/2002

<OWNER INFORMATION>

<Name>   MARTIN SAND
<Street>   123 DESERT PARKWAY
<City>   MOAB
<State>   UTAH
<Country> USA
<Zip/Postal Code>   80123

<DOMESTIC REPRESENTATIVE>~*Required ONLY if the owner's address is outside the United States.*~

<Name>   ~is hereby appointed the owner's representative
upon whom notice or process in the proceedings affecting the mark may be served.~

<Street>

<City>

<State>

<Zip Code>

<GOODS AND/OR SERVICES INFORMATION>

<All Goods and/or Services in Existing Registration>~*The owner is using the mark in commerce on or in connection with all goods and/or services listed in the existing registration. If not, list in the next section the goods and/or services to be deleted.*~

<Goods and/or Services Not in Use to be **Deleted**>~*In the following space, list only those goods and/or services (or entire classes(es)) appearing in the registration for which the owner is **no longer** using the mark in commerce. **LEAVE THIS SPACE BLANK IF THE OWNER IS USING THE MARK ON OR IN CONNECTION WITH ALL GOODS AND/OR SERVICES LISTED IN THE REGISTRATION.**~*

ONLY FILL IN THIS SECTION IF YOU ARE NO LONGER USING THE MARK ON SOME OF THE GOODS FOR WHICH IT IS REGISTERED

<FEE INFORMATION>

~*Section 8 Filing Fee*~
$100.00 x <Number of Classes>   1   = <Filing Fee Due> $100.00
~*Grace Period Fee: If filing during the six-month grace period, enter § 8 Grace Period Fee*~
$100.00 x <Number of Classes>   = <Grace Period Fee Due>
   ~*Filing Fee Due + Grace Period Fee Due*~ = <Total Fees Paid> $100.00

ONLY PAY THE GRACE PERIOD FEE IF YOU ARE FILING LATE

PTO Form 1583 (REV 5/99)
OMB Control No. 0651-0009 (Exp. 8/31/2001)

U.S. DEPARTMENT OF COMMERCE/Patent and Trademark Office
There is no requirement to respond to this collection of information
unless a currently valid OMB number is displayed.

~DECLARATION OF INCONTESTABILITY OF MARK UNDER §15 (15 U.S.C. § 1065)~

~To the Assistant Commissioner for Trademarks~

<TRADEMARK/SERVICE MARK INFORMATION>

<Mark>	SAHARA
<Registration Number>	1357924680
<Registration Date>	02/02/2002

<OWNER INFORMATION>

<Name>	MARTIN SAND
<Street>	123 DESERT PARKWAY
<City>	MOAB
<State>	UTAH
<Country>	USA
<Zip/Postal Code>	80123

<GOODS AND/OR SERVICES INFORMATION>

<All Goods and/or Services in Existing Registration>~*The owner has used the mark in commerce for five (5) consecutive years after the date of registration, or the date of publication under § 12(c), and is still using the mark in commerce on or in connection with all goods and/or services listed in the existing registration. If not, list in the next section the goods and/or services not covered.*~

<Goods and/or Services **Not Covered**>~*In the following space, list only those goods and/or services (or entire classes(es)) appearing in the registration for which either the owner has NOT used the mark in commerce for five (5) consecutive years or is NO LONGER using the mark in commerce.* **LEAVE THIS SPACE BLANK IF THE OWNER HAS USED THE MARK IN COMMERCE FOR FIVE (5) CONSECUTIVE YEARS AFTER THE DATE OF REGISTRATION, OR THE DATE OF PUBLICATION UNDER § 12(C), AND IS STILL USING THE MARK IN COMMERCE ON OR IN CONNECTION WITH ALL GOODS/ SERVICES IN THE EXISTING REGISTRATION.*~

ONLY FILL IN THIS SECTION IF YOU ARE NO LONGER USING THE MARK ON SOME OF THE GOODS FOR WHICH IT IS REGISTERED

PTO Form 4.16 (REV 5/99)
OMB Control No. 0651-0009 (Exp. 8/31/2001)

U.S. DEPARTMENT OF COMMERCE/Patent and Trademark Office
There is no requirement to respond to this collection of information
unless a currently valid OMB number is displayed.

145

~DECLARATION OF USE OF MARK IN COMMERCE UNDER § 8 (15 U.S.C. § 1058)~

~To the Assistant Commissioner for Trademarks~

<TRADEMARK/SERVICE MARK INFORMATION>

<Mark>   SAHARA
<Registration Number> 1357924680
<Registration Date>   02/02/2002

<OWNER INFORMATION>

<Name> MARTIN SAND
<Street> 123 DESERT PARKWAY
<City>   MOAB
<State>   UTAH
<Country>USA
<Zip/Postal Code> 80123

<DOMESTIC REPRESENTATIVE>~*Required ONLY if the owner's address is outside the United States.*~

<Name>   ~is hereby appointed the owner's representative upon whom notice or process in the proceedings affecting the mark may be served.~

<Street>

<City>

<State>

<Zip Code>

<GOODS AND/OR SERVICES INFORMATION>

<All Goods and/or Services in Existing Registration>~*The owner is using the mark in commerce on or in connection with all goods and/or services listed in the existing registration. If not, list in the next section the goods and/or services to be deleted.*~

<Goods and/or Services Not in Use to be **Deleted**>~*In the following space, list only those goods and/or services (or entire classes(es)) appearing in the registration for which the owner is **no longer** using the mark in commerce. **LEAVE THIS SPACE BLANK IF THE OWNER IS USING THE MARK ON OR IN CONNECTION WITH ALL GOODS AND/OR SERVICES LISTED IN THE REGISTRATION.***~

*ONLY FILL IN THIS SECTION
IF YOU ARE NO LONGER
USING THE MARK ON SOME
OF THE GOODS FOR
WHICH IT IS REGISTERED*

<FEE INFORMATION>

~*Section 8 Filing Fee*~
$100.00 x <Number of Classes>   1   = <Filing Fee Due> $100.00
~*Grace Period Fee: If filing during the six-month grace period, enter § 8 Grace Period Fee*~
$100.00 x <Number of Classes>   = <Grace Period Fee Due>
      ~*Filing Fee Due + Grace Period Fee Due*~ = <Total Fees Paid> $100.00

*ONLY PAY THE GRACE
PERIOD FEE IF YOU
ARE FILING LATE*

PTO Form 1583 (REV 5/99)
OMB Control No. 0651-0009 (Exp. 8/31/2001)

U.S. DEPARTMENT OF COMMERCE/Patent and Trademark Office
There is no requirement to respond to this collection of information
unless a currently valid OMB number is displayed.

146

COMBINED DECLARATION OF USE IN COMMERCE/APPLICATION FOR RENEWAL OF REGISTRATION OF MARK UNDER §§ 8 & 9 (15 U.S.C. §§ 1058 & 1059)~

~To the Assistant Commissioner for Trademarks~

\<TRADEMARK/SERVICE MARK INFORMATION>

\<Mark>   SAHARA
\<Registration Number> 1357924680
\<Registration Date>   02/02/2002

\<OWNER INFORMATION>

\<Name>   MARTIN SAND
\<Street>   123 DESERT PARKWAY
\<City>   MOAB
\<State>   UTAH
\<Country> USA
\<Zip/Postal Code>   80123

\<DOMESTIC REPRESENTATIVE>~*Required ONLY if the owner's address is outside the United States.*~

\<Name>   ~is hereby appointed the owner's
representative upon whom notice or process in the proceedings affecting the mark may be served.~

\<Street>
\<City>
\<State>
\<Zip Code>

\<GOODS AND/OR SERVICES INFORMATION>

\<All Goods and/or Services in Existing Registration>~*The owner is using mark in commerce on or in connection with all goods and/or services listed in the existing registration. If not, list in the next section the goods and/or services to be deleted.*~

\<Goods and/or Services Not in Use to be **Deleted**>~*In the following space, list only those goods and/or services (or entire classes(es)) appearing in the registration for which the owner is **no longer** using the mark in commerce. **LEAVE THIS SPACE BLANK IF THE OWNER IS USING THE MARK ON OR IN CONNECTION WITH ALL GOODS AND/OR SERVICES LISTED IN THE REGISTRATION).***~

> ONLY FILL IN THIS SECTION IF YOU ARE NO LONGER USING THE MARK ON SOME OF THE GOODS FOR WHICH IT IS REGISTERED

\<FEE INFORMATION>

~*Combined §§ 8 & 9 Filing Fee*~
$400.00 x \<Number of Classes>   1 = \<Filing Fee Due> $400.00
~*Grace Period Fee: If filing during the six-month grace period, enter Combined §§ 8 & 9 Grace Period Fee.*~
$200.00 x \<Number of Classes>   = \<Grace Fee Due>
~*Filing Fee Due + Grace Period Fee Due*~ = \<Total Fees Paid> $400.00

> ONLY PAY THE GRACE PERIOD FEE IF YOU ARE FILING LATE

U.S. DEPARTMENT OF COMMERCE/Patent and Trademark Office
There is no requirement to respond to this collection of information
unless a currently valid OMB number is displayed.

DESIGNATION OF DOMESTIC REPRESENTATIVE	MARK *(identify the mark)* PERIODICO UNO
	REGISTRATION NO. (IF KNOWN) 23456789
	CLASS NO. (S)

WILLIAM D. VAN PIPPERMORE III

(name of domestic representative)

whose postal address is <u>1001 AVENUE OF THE ATTORNEYS, NEW YORK, NY 10001</u>

_____ is hereby designated applicant's representative upon whom notice or process in proceedings affecting the mark may be served.

William D. Van Pippermore III

(signature of applicant or owner of mark)

JULY 4, 1999

(date)

	MARK *(identify the mark)*
	M+M
ASSIGNMENT OF	REGISTRATION NO. (IF KNOWN)
REGISTRATION OF A MARK	12345678
	CLASS NO. (S)
	30

Whereas_____ MARY MAJORS _____
(name of assignor)

whose postal address is ___ 123 CENTRAL AVE., WESTPORT ME 01212 _____

_____ has adopted, used and is using a mark which is registered in the

United States Patent and Trademark Office, Registration No. ___ 12345678 _____ dated

JULY 17, 1973 _____; and whereas ___ MARCIA NOVAK _____
(name of assignee)

whose postal address is _456 DUHME RD. FALLS CHURCH, VA 22044_____

is desirous of acquiring said mark and the registration thereof;

Now, therefore, for good and valuable consideration, receipt of which is hereby acknowledged, said

___ MARY MAJORS _____ does hereby assign unto the said
(name of assignor)

___ MARCIA NOVAK _____ all right, title and interest in and to
(name of assignee)

the said mark, together with the good will of the business symbolized by the mark, and the above registration

thereof.

_____ *MARY MAJORS* _____
*(signature of assignor, if assignor is a corporation or other juristic organization give
the official title of the person who signs for assignor)*

State of ___ MAINE _____ ⎫
County of _ LEE _____ ⎬ ss.
⎭

On this _1__ day of AUGUST _____, 19_98_, before me appeared __ MARY MAJORS _____

_____ the person who signed this instrument, who acknowledged that he/she

signed it as a free act on his/her own behalf (or on behalf of the identified corporation or other juristic entity

with authority to do so).*

_____ *C.U. Sine* _____
(signature of notary public)

* The wording of the acknowledgment may vary in some jurisdictions. Be sure to use wording acceptable in the jurisdiction where the document is executed.

FORM PTO-1618A
Expires 06/30/99
OMB 0651-0027

U.S. Department of Commerce
Patent and Trademark Office
TRADEMARK

RECORDATION FORM COVER SHEET
TRADEMARKS ONLY

TO: The Commissioner of Patents and Trademarks: Please record the attached original document(s) or copy(ies).

Submission Type

[X] **New**

[ ] **Resubmission** **(Non-Recordation)**
Document ID #

[ ] **Correction of PTO Error**
Reel #   Frame #

[ ] **Corrective Document**
Reel #   Frame #

Conveyance Type

[X] **Assignment**   [ ] **License**

[ ] **Security Agreement**   [ ] **Nunc Pro Tunc Assignment**

[ ] **Merger**

Effective Date
Month Day Year
06 05 99

[ ] **Change of Name**

[ ] **Other**

Conveying Party

[ ] Mark if additional names of conveying parties attached

Execution Date
Month Day Year
06 05 99

Name   SPIDER WEBBING, INC.

Formerly

[ ] **Individual**   [ ] **General Partnership**   [ ] **Limited Partnership**   [X] **Corporation**   [ ] **Association**

[ ] **Other**

[ ] **Citizenship/State of Incorporation/Organization**   DISTRICT OF COLUMBIA

Receiving Party

[ ] Mark if additional names of receiving parties attached

Name   WEB MASTER, INC.

DBA/AKA/TA

Composed of

Address (line 1)   123 MAIN ST.

Address (line 2)

Address (line 3)   NAPERVILLE   IL   60540
      City   State/Country   Zip Code

[ ] **Individual**   [ ] **General Partnership**   [ ] **Limited Partnership**

[X] **Corporation**   [ ] **Association**

[ ] **Other**

If document to be recorded is an assignment and the receiving party is not domiciled in the United States, an appointment of a domestic representative should be attached. *(Designation must be a separate document from Assignment.)*

[ ] **Citizenship/State of Incorporation/Organization**   ILLINOIS

FOR OFFICE USE ONLY

Public burden reporting for this collection of information is estimated to average approximately 30 minutes per Cover Sheet to be recorded, including time for reviewing the document and gathering the data needed to complete the Cover Sheet. Send comments regarding this burden estimate to the U.S. Patent and Trademark Office, Chief Information Officer, Washington, D.C. 20231 and to the Office of Information and Regulatory Affairs, Office of Management and Budget, Paperwork Reduction Project (0651-0027), Washington, D.C. 20503. See OMB Information Collection Budget Package 0651-0027, Patent and Trademark Assignment Practice. DO NOT SEND REQUESTS TO RECORD ASSIGNMENT DOCUMENTS TO THIS ADDRESS.

Mail documents to be recorded with required cover sheet(s) information to:
Commissioner of Patents and Trademarks, Box Assignments , Washington, D.C. 20231

FORM PTO-1618B
Expires 06/30/99
OMB 0651-0027

Page 2

U.S. Department of Commerce
Patent and Trademark Office
TRADEMARK

Domestic Representative Name and Address   Enter for the first Receiving Party only.

Name

Address (line 1)

Address (line 2)

Address (line 3)

Address (line 4)

Correspondent Name and Address   Area Code and Telephone Number

Name

Address (line 1)

Address (line 2)

Address (line 3)

Address (line 4)

Pages   Enter the total number of pages of the attached conveyance document including any attachments.

Trademark Application Number(s) or Registration Number(s)   [X] Mark if additional numbers attached

Enter either the Trademark Application Number or the Registration Number (DO NOT ENTER BOTH numbers for the same property).

Trademark Application Number(s)			Registration Number(s)		
			1234567	4567890	7890123
			2345678	5678901	8901234
			3456789	6789012	9012345

Number of Properties   Enter the total number of properties involved.   # 10

Fee Amount   Fee Amount for Properties Listed (37 CFR 3.41): $ 1,000.00

Method of Payment: Enclosed [X]   Deposit Account [ ]

Deposit Account
(Enter for payment by deposit account or if additional fees can be charged to the account.)
Deposit Account Number: #

Authorization to charge additional fees: Yes [X]   No [ ]

Statement and Signature

To the best of my knowledge and belief, the foregoing information is true and correct and any attached copy is a true copy of the original document. Charges to deposit account are authorized, as indicated herein.

Boris Spider	*Boris Spider*	JUNE 5, 1999
Name of Person Signing	**Signature**	**Date Signed**

151

RECORDATION FORM COVER SHEET
CONTINUATION
TRADEMARKS ONLY

FORM PTO-1618C
Expires 06/30/99
OMB 0651-0027

U.S. Department of Commerce
Patent and Trademark Office
TRADEMARK

Conveying Party
Enter Additional Conveying Party

☐ Mark if additional names of conveying parties attached

Execution Date
Month Day Year
06 05 99

Name: SPIDER WEBBING, INC.

Formerly: [ ]

☐ Individual   ☐ General Partnership   ☐ Limited Partnership   ☒ Corporation   ☐ Association

☐ Other [ ]

☐ Citizenship State of Incorporation/Organization DISTRICT OF COLUMBIA

Receiving Party
Enter Additional Receiving Party

☐ Mark if additional names of receiving parties attached

Name: WEB MASTER, INC.

DBA/AKA/TA: [ ]

Composed of: [ ]

Address (line 1): 123 MAIN ST.

Address (line 2): [ ]

Address (line 3): NAPERVILLE | IL | 60540
City | State/Country | Zip Code

☐ Individual   ☐ General Partnership   ☐ Limited Partnership

☒ Corporation   ☐ Association

☐ If document to be recorded is an assignment and the receiving party is not domiciled in the United States, an appointment of a domestic representative should be attached *(Designation must be a separate document from the Assignment.)*

☐ Other [ ]

☐ Citizenship/State of Incorporation/Organization [ ]

Trademark Application Number(s) or Registration Number(s)

☐ Mark if additional numbers attached

Enter either the Trademark Application Number or the Registration Number (DO NOT ENTER BOTH numbers for the same property).

Trademark Application Number(s)			Registration Number(s)		
			0123456		

Application to Record Trademark
with the United States Customs Service

To: Intellectual Property Rights Branch
      U. S. Customs Service
      1301 Constitution Ave., N.W.
      Washington, DC 20229

Name of trademark owner: ATOM INDUSTRIES, INC

Address of trademark owner: 102 SUNSET BLVD.
                             KEY WEST FLORIDA 33410

Trademark owner is:
☐ an individual who is a citizen of _____
☐ a partnership whose partners are citizens of _____
☒ an association or corporation which was organized under the laws of
       NEVADA_____

Places of manufacture of goods bearing the trademark:

     FLORIDA, USA

The following are foreign persons authorized to use the trademark:

Name: Address: Use authorized:

       NONE

Identification of any foreign parent or subsidiaries under common ownership or control
which uses the trademark abroad*:

       NONE

Include with this form:
1. A status copy of the certificate of registration certified by the U. S. Patent and
    Trademark Office showing title to be presently in the name of the applicant.
2. Five copies of the certificate or of a U. S. Patent and Trademark Office Certificate.
3. A fee of $190 for each class of goods sought to be protected.

*Note, "common ownership" means individual or aggregate ownership of more than 50% of the business entity and
"common control" means effective control in policy and operations and is not necessarily synonymous with common
ownership.

Appendix C
Blank Forms
& Instructions

The following pages contain the forms mentioned throughout this book. They may be either removed from this book and used or photocopied. The forms are occasionally revised by the Patent and Trademark Office, but in most cases the previous versions of the forms are acceptable. If you would like to see if new versions of the forms are available, you can check the Patent and Trademark Office website at http://www.uspto.gov/web/forms/.

_____

[Your Name]

_____

[Address]

_____

[City, State Zip]

_____

[Phone]

[Date]

Hon Commissioner of Patents and Trademarks
Washington, DC 20231

Re: In Use Trademark Application

Sir:

Enclosed for filing please find trademark application papers of _____
_____ for the mark "_____" in International
Class _____ together with the drawing, a specimen and the filing fee of $325.

A self-addressed, stamped envelope is enclosed for return of the receipt.

Respectfully submitted,

_____

[Your Name]

_____

[Your Name]

_____

[Address]

_____

[City, State Zip]

_____

[Phone]

[Date]

Hon Commissioner of Patents and Trademarks
Washington, DC 20231

Re: Intent-to-Use Trademark Application

Sir:

Enclosed for filing please find trademark application papers of _____
_____ for the mark "_____" in International
Class _____ together with the drawing and the filing fee of $325.

A self-addressed, stamped envelope is enclosed for return of the receipt.

Respectfully submitted,

_____

[Your Name]

~TRADEMARK/SERVICE MARK APPLICATION (15 U.S.C. §§ 1051, 1126(d)&(e))~

NOTE: The following form complies with the provisions of the Trademark Law Treaty Implementation Act (TLTIA).

BASIC INSTRUCTIONS

The following form is written in a "scannable" format that will enable the U.S. Patent and Trademark Office (USPTO) to scan paper filings and capture application data automatically using optical character recognition (OCR) technology. Information is to be entered next to identifying data tags, such as <DATE OF FIRST USE IN COMMERCE>. OCR software can be programmed to identify these tags, capture the corresponding data, and transmit this data to the appropriate data fields in the Trademark databases, largely bypassing manual data entry processes.

Please enter the requested information in the blank space that appears to the right of each tagged (< >) element. However, do not enter any information immediately after the section headers (the bolded wording appearing in all capital letters). If you need additional space, first, in the space provided on the form, enter "See attached." Then, please use a separate piece of paper on which you first list the data tag (e.g., <LISTING OF GOODS AND/OR SERVICES>), followed by the relevant information. Some of the information requested *must* be provided. Other information is either required only in certain circumstances, or provided only at your discretion. **Please consult the "Help" section following the form for detailed explanations as to what information should be entered in each blank space.**

To increase the effectiveness of the USPTO scanners, it is recommended that you use a typewriter to complete the form.

For additional information, please see the *Basic Facts about Trademarks* booklet, available at http://www.uspto.gov/web/offices/tac/doc/basic/, or by calling the Trademark Assistance Center, at 703-308-9000. You may also wish to file electronically, from http://www.uspto.gov/teas/index.html.

MAILING INFORMATION

Send the completed form, appropriate fee(s) (made payable to "The Commissioner of Patent and Trademarks"), and any other required materials to:

> Box New App
> Fee
> Assistant Commissioner for Trademarks
> 2900 Crystal Drive
> Arlington, VA 22202-3513

The filing fee for this application is $245.00 *per class* of goods and/or services. You must include at least $245.00 with this application; otherwise the papers and money will be returned to you. Once your application meets the minimum filing date requirements, this processing fee becomes **non-refundable**. This is true even if the USPTO does not issue a registration certificate for this mark.

You may also wish to include a self-addressed stamped postcard with your submission, on which you identify the mark and list each item being submitted (e.g., application, fee, specimen, etc.). We will return this postcard to you, stamped with your assigned serial number, to confirm receipt of your submission.

~TRADEMARK/SERVICE MARK APPLICATION (15 U.S.C. §§ 1051, 1126(d)&(e))~

~To the Assistant Commissioner for Trademarks~

<APPLICANT INFORMATION>

<Name>

<Street>

<City>

<State>

<Country>

<Zip/Postal Code>

<Telephone Number>

<Fax Number>

<e-mail Address>

<APPLICANT ENTITY INFORMATION>~Select only ONE~

<Individual: Country of Citizenship>

<Corporation: State/Country of Incorporation>

<Partnership: State/Country under which Organized>

<Name(s) of General Partner(s) & Citizenship/Incorporation>

<Other Entity Type: Specific Nature of Entity>

<State/Country under which Organized>

<TRADEMARK/SERVICE MARK INFORMATION>

<Mark>

<Typed Form>~*Enter YES, if appropriate*~

~*DISPLAY THE MARK that you want to register on a separate piece of paper (even if simply a word(s)). Please see additional HELP instructions.*~

<BASIS FOR FILING AND GOODS/SERVICES INFORMATION>

<**Use in Commerce: Section 1(a)**>~*Applicant is using the mark in commerce on or in connection with the below-identified goods and/or services (15 U.S.C § 1051(a)).*~

<International Class Number(s)>

<Listing of Goods and/or Services>~**List in ascending numerical class order. Please see sample in HELP instructions.**~

<Date of First Use Anywhere>

<Date of First Use in Commerce>

~**Submit one (1) SPECIMEN for each international class showing the mark as used in commerce.**~

<**Intent to Use: Section 1(b)**>~*Applicant has a bona fide intention to use the mark in commerce on or in connection with the below-identified goods and/or services (15 U.S.C. § 1051(b)).*~

<International Class Number(s)>

<Listing of Goods and/or Services>~***List in ascending numerical class order. Please see sample in HELP instructions.***~

<**Foreign Priority: Section 44(d)**>~*Applicant has a bona fide intention to use the mark in commerce on or in connection with the below-identified goods and/or services, and asserts a claim of priority based upon a foreign application in accordance with 15 U.S.C. § 1126(d).*~

<International Class Number(s)>

<Listing of Goods and/or Services>~***List in ascending numerical class order. Please see sample in HELP instructions.***~

<Country of Foreign Filing>

<Foreign Application Number>

<Date of Foreign Filing>

<**Foreign Registration: Section 44(e)**>~*Applicant has a bona fide intention to use the mark in commerce on or in connection with the below-identified goods and/or services based on registration of the mark in applicant's country of origin.*~

<International Class Number(s)>

<Listing of Goods and/or Services>~***List in ascending numerical class order. Please see sample in HELP instructions.***~

<Country of Foreign Registration>

<Foreign Registration Number>

<Foreign Registration Date>

<Foreign Registration Renewal Date>

<Foreign Registration Expiration Date>

~***Submit foreign registration certificate or a certified copy of the foreign registration, in accordance with 15 U.S.C. §1126(e).***~

<FEE INFORMATION>

$245.00 x <Number of Classes>	= <Total Filing Fee Paid>	

< SIGNATURE INFORMATION>

~Applicant requests registration of the above-identified mark in the United States Patent and Trademark Office on the Principal Register established by Act of July 5, 1946 (15 U.S.C. § 1051 et seq.) for the above-identified goods and/or services.

The undersigned, being hereby warned that willful false statements and the like so made are punishable by fine or imprisonment, or both, under 18 U.S.C. § 1001, and that such willful false statements may jeopardize the validity of the application or any resulting registration, declares that he/she is properly authorized to execute this application on behalf of the applicant; he/she believes the applicant to be the owner of the trademark/service mark sought to be registered, or, if the application is being filed under 15 U.S.C. § 1051(b), he/she believes applicant to be entitled to use such mark in commerce; to the best of his/her knowledge and belief no other person, firm, corporation, or association has the right to use the mark in commerce, either in the identical form thereof or in such near resemblance thereto as to be likely, when used on or in connection with the goods/services of such other person, to cause confusion, or to cause mistake, or to deceive; and that all statements made of his/her own knowledge are true; and that all statements made on information and belief are believed to be true.~

~Signature~_____

<Date>

<Name>

<Title>

<CONTACT INFORMATION>

<Name>

<Company/Firm Name>

<Street>

<City>

<State>

<Country>

<Zip/Postal Code>

<Telephone Number>

<Fax Number>

<e-Mail Address>

LINE-BY-LINE HELP INSTRUCTIONS

APPLICANT INFORMATION

Name: Enter the full name of the owner of the mark, i.e., the name of the individual, corporation, partnership, or other entity that is seeking registration. If a joint venture organized under a particular business name owns the mark, enter that name. If joint or multiple owners, enter the name of each of these owners. If a trust, enter the name of the trustee(s). If an estate, enter the name of the executor(s).

Street: Enter the street address or rural delivery route where the applicant is located.

City: Enter the city and/or foreign area designation where the applicant's address is located.

State: Enter the U.S. state or foreign province in which the applicant's address is located.

Country: Enter the country of the applicant's address. If the address is outside the United States, the applicant must appoint a "Domestic Representative" on whom notices or process in proceedings affecting the mark may be served.

Zip/Postal Code: Enter the applicant's U.S. Zip code or foreign country postal identification code.

Telephone Number: Enter the applicant's telephone number.

Fax Number: Enter the applicant's fax number.

e-Mail Address: Enter the applicant's e-mail address.

APPLICANT ENTITY INFORMATION

Indicate the applicant's entity type by entering the appropriate information in the space to the right of the correct entity type. Please note that only one entity type may be selected.

Individual: Enter the applicant's country of citizenship.

Corporation: Enter the applicant's state of incorporation (or the applicant's country of incorporation if the applicant is a foreign corporation).

Partnership: Enter the state under whose laws the partnership is organized (or the country under whose laws the partnership is organized if the partnership is a foreign partnership).

Name(s) of General Partner(s) & Citizenship/incorporation: Enter the names and citizenship of any general partners who are individuals, and/or the names and state or (foreign) country of incorporation of any general partners that are corporations, and/or the names and states or (foreign) countries of organization of any general partners that are themselves partnerships. If the owner is a limited partnership, then only provide the names and citizenship or state or country of organization or incorporation of the general partners.

Other Entity Type: Enter a brief description of the applicant's entity type (e.g., joint or multiple applicants, joint venture, limited liability company, association, Indian Nation, state or local agency, trust, estate). The following sets forth the information required with respect to the most common types of "other" entities:

For *joint or multiple applicants*, enter the name and entity type of each joint applicant. Also, enter the citizenship of those joint applicants who are individuals, and/or the state or (foreign) country of incorporation of those joint applicants that are corporations, and/or the state or (foreign) country of organization- and the names and citizenship of the partners- of those joint applicants that are partnerships. The information regarding each applicant should be preceded by a separate heading tag (<APPLICANT INFORMATION>).

For *sole proprietorship,* enter the name and citizenship of the sole proprietor, and indicate the state where the sole proprietorship is organized.

For *joint venture*, enter the name and entity type of each entity participating in the joint venture. Also, enter the citizenship of those joint venture participants who are individuals, and/or the state or (foreign) country of incorporation of those joint venture participants that are corporations, and/or the state or (foreign) country of organization (and the names and citizenship of the partners) of those joint venture participants that are partnerships. The information regarding each entity should be preceded by a separate heading tag (<APPLICANT INFORMATION>).

For *limited liability company or association*, enter the state or (foreign) country under whose laws the entity is established.

165

For *state or local agency*, enter the name of the agency and the state and/or locale of the agency (e.g., Maryland State Lottery Agency, an agency of the State of Maryland).

For *trusts*, identify the trustees and the trust itself, using the following format: The Trustees of the XYZ Trust, a California trust, the trustees comprising John Doe, a U.S. citizen, and the ABC Corp., a Delaware corporation. (Please note that the trustees, and not the trust itself, must be identified as the applicant in the portion of the application designated for naming the applicant).

For *estates*, identify the executors and the estate itself using the following format: The Executors of the John Smith estate, a New York estate, the executors comprising Mary Smith and John Smith, U.S. citizens. (Please note that the executors, and not the estate itself, must be identified as the applicant in the portion of the application designated for naming the applicant).

State/Country under Which Organized: Enter the state or country under whose laws the entity is organized.

TRADEMARK/SERVICE MARK INFORMATION

Mark: A mark may consist of words alone; a design or logo; or a combination of words and a design or logo. However, an application may consist of only *one* mark; separate marks must be filed in separate applications. In this space, enter the word mark in typed form (e.g., THE CAT'S MEOW); or, in the case of a non-word mark or a combination mark, a brief description of the mark (e.g., Design of a fanciful cat or Design of a fanciful cat and the words THE CAT'S MEOW). Do NOT include quotation marks around the mark itself, unless the mark actually features these quotation marks. Also, do NOT include any information related to a "pseudo mark" in this field, because only the USPTO controls this field. If the USPTO determines that a pseudo mark is necessary for your particular mark, it will enter this information in the search system.

Typed Form: Enter YES if the mark applied for is in a "typed" format (i.e., if the mark consists of *only* typed words, letters or numbers, and does *not* include any special stylization or design element(s)). Please note that a registration for a mark based on a typed drawing affords protection not only for the typed version of the mark, but for all other renderings of the mark as long as those renderings do not contain any design elements.

Display the Mark: Regardless of whether the mark consists of words alone; a design or logo; or a combination of words and a design or logo, submit on a separate piece of paper a display of what the mark is.

At the top of the page, include a heading consisting of (1) the applicant's name and address; (2) a listing of the goods and/or services on which on in connection with which the mark is used; and (3) a listing of the basis for filing (and any relevant information related thereto). Then, in the middle of the page, show the mark:

If the mark is to be in a "typed" form, simply type the mark in the middle of the page *in all capital letters*. For a mark in stylized form or design, in the middle of the page display an image of the mark in black and white, in an area no greater than 4x4 inches.

BASIS FOR FILING AND GOODS/SERVICES INFORMATION

Use in Commerce: Section 1(a): Use this section only if you have actually used the mark in commerce or on in connection with *all* of the goods and/or services listed.

International Class Number(s): Enter the International Class number(s) of the goods and/or services associated with the mark; e.g., 14; 24; 25. If unknown, leave blank and the USPTO will assign the number(s).

Listing of Goods and/or Services: Enter the *specific* goods and/or services associated with the mark. Do NOT enter the broad class number here, such as 9 or 42 (this information belongs in the field above, namely International Class Number(s)). If the goods and/or services are classified in more than one class, the goods and/or services should be listed in ascending numerical class order, with both the class number and the specific goods and/or services. For example, 14: jewelry

24: towels

25: pants, shirts, jackets, shoes

For more information about acceptable wording for the goods/services, see the USPTO's on-line *Acceptable Identification of Goods and Services Manual*, at http://www.uspto.gov/web/offices/tac/doc/gsmanual/.

Date of First Use Anywhere: Enter the date on which the goods were first sold or transported or the services first rendered under the mark if such use was in the ordinary course of trade. For every applicant (foreign or domestic), the date of first use is the date of the first such use *anywhere*, in the United States or elsewhere. Please note this date may be earlier than, or the same as, the date of the first use of the mark in commerce.

Date of First Use in Commerce: Enter the date on which the applicant first used the mark in commerce, i.e., in interstate commerce, territorial commerce, or commerce between the United States and a foreign country.

Specimen: You must submit one (1) specimen showing the mark as used in commerce on or in connection with any item listed in the description of goods and/or services; e.g., tags or labels for goods, and/or advertisements for services. If the goods and/or services are classified in more than one international class, a specimen must be provided showing the mark used on or in connection with at least one item from each of these classes. The specimen must be flat and no larger than 8½ inches (21.6 cm.) wide by 11.69 inches (29.7 cm.) long.

Intent to Use: Section 1(b): Use this section if the applicant only has a bona fide intention to use the mark in commerce in the future as to all or some of the goods and/or services, rather than having actually already made use of the mark in commerce as to *all* of the goods and/or services.

International Class Number(s): Enter the International Class number(s) of the goods and/or services associated with the mark; e.g., 14; 24; 25. If unknown, leave blank and the USPTO will assign the number(s).

Listing of Goods and/or Services: Enter the *specific* goods and/or services associated with the mark. Do NOT enter the broad class number here, such as 9 or 42 (this information belongs in the field above, namely International Class Number(s)). If the goods and/or services are classified in more than one class, the goods and/or services should be listed in ascending numerical class order, with both the class number and the specific goods and/or services. For example, 14: jewelry

              24: towels

              25: pants, shirts, jackets, shoes

For more information about acceptable wording for the goods/services, see the USPTO's on-line *Acceptable Identification of Goods and Services Manual*, at http://www.uspto.gov/web/offices/tac/doc/gsmanual/.

Foreign Priority: Section 44(d): Use this section if you are filing the application within six (6) months of filing the first foreign application to register the mark in a defined treaty country.

International Class Number(s): Enter the International Class number(s) of the goods and/or services associated with the mark; e.g., 14; 24; 25. If unknown, leave blank and the USPTO will assign the number(s).

Listing of Goods and/or Services: Enter the *specific* goods and/or services associated with the mark. Do NOT enter the broad class number here, such as 9 or 42 (this information belongs in the field above, namely International Class Number(s)). If the goods and/or services are classified in more than one class, the goods and/or services should be listed in ascending numerical class order, with both the class number and the specific goods and/or services. For example, 14: jewelry

              24: towels

              25: pants, shirts, jackets, shoes

For more information about acceptable wording for the goods/services, see the USPTO's on-line *Acceptable Identification of Goods and Services Manual*, at http://www.uspto.gov/web/offices/tac/doc/gsmanual/.

Country of Foreign Filing: Enter the country where the foreign application upon which the applicant is asserting a claim of priority has been filed.

Foreign Application Number: Enter the foreign application serial number, if available.

Filing Date of Foreign Application: Enter the date (two digits each for both the month and day, and four digits for the year) on which the foreign application was filed. To receive a priority filing date, you must file the U.S. application within six (6) months of filing the first foreign application in a defined treaty country.

Foreign Registration: Use this section if applicant is relying on a foreign registration certificate or a certified copy of a foreign registration currently in force. You must submit this foreign registration certificate or a certified copy of the foreign registration.

International Class Number(s): Enter the International Class number(s) of the goods and/or services associated with the mark; e.g., 14; 24; 25. If unknown, leave blank and the USPTO will assign the number(s).

Listing of Goods and/or Services: Enter the *specific* goods and/or services associated with the mark. Do NOT enter the broad class number here, such as 9 or 42 (this information belongs in the field above, namely **167**

International Class Number(s)). If the goods and/or services are classified in more than one class, the goods and/or services should be listed in ascending numerical class order, with both the class number and the specific goods and/or services. For example, 14: jewelry

24: towels

25: pants, shirts, jackets, shoes

For more information about acceptable wording for the goods/services, see the USPTO's on-line *Acceptable Identification of Goods and Services Manual*, at http://www.uspto.gov/web/offices/tac/doc/gsmanual/.

Country of Foreign Registration: Enter the country of the foreign registration.

Foreign Registration Number: Enter the number of the foreign registration.

Foreign Registration Date: Enter the date (two digits each for both the month and day, and four digits for the year) of the foreign registration.

Foreign Registration Renewal Date: Enter the date (two digits each for both the month and day, and four digits for the year) of the foreign registration renewal.

Foreign Registration Expiration Date: Enter the expiration date (two digits each for both the month and day, and four digits for the year) of the foreign registration.

FEE INFORMATION

The filing fee for this application is $245.00 *per class* of goods and/or services. You must include at least $245.00 with this application; otherwise the papers and money will be returned to you. Once your application meets the minimum filing date requirements, this processing fee becomes **non-refundable**. This is true even if the USPTO does not issue a registration certificate for this mark.

Number of Classes: Enter the total number of classes (*not* the international class number(s)) for which the applicant is seeking registration. For example, if the application covers Classes 1, 5 and 25, then enter the number "3."

Total Filing Fee Paid: Enter the fee amount that is enclosed (either in the form of a check or money order in U.S. currency, made payable to "Commissioner of Patents and Trademarks"), or to be charged to an already-existing USPTO deposit account.

SIGNATURE INFORMATION

Signature: The appropriate person must sign the form. A person who is properly authorized to sign on behalf of the owner is: (1) a person with legal authority to bind the owner; or (2) a person with firsthand knowledge of the facts and actual or implied authority to act on behalf of the owner; or (3) an attorney who has an actual or implied written or verbal power of attorney from the owner.

Date Signed: Enter the date the form is signed.

Name: Enter the name of the person signing the form.

Title: Enter the signatory's title, if applicable, e.g., Vice-President, General Partner, etc.

CONTACT INFORMATION

Although this may be the same as provided elsewhere in the document, please enter the following required information for where the USPTO should mail correspondence. (Please note that correspondence will *only* be mailed to an address in the U.S. or Canada).

Name: Enter the full name of the contact person.

Company/Firm Name: Enter the name of the contact person's company or firm.

Street: Enter the street address or rural delivery route where the contact person is located.

City: Enter the city and/or foreign area designation where the contact person's address is located.

State: Enter the U.S. state or Canadian province in which the contact person's address is located.

Country: Enter the country of the contact person's address.

Zip Code: Enter the U.S. Zip code or Canadian postal code.

Telephone Number: Enter the appropriate telephone number.

Fax Number: Enter the appropriate fax number, if available.

e-mail Address: Enter the appropriate e-mail address, if available.

COLLECTIVE TRADEMARK/SERVICE MARK APPLICATION, PRINCIPAL REGISTER, WITH DECLARATION	MARK (Word(s) and/or Design)	CLASS NO. (If known)

TO THE ASSISTANT COMMISSIONER FOR TRADEMARKS:

APPLICANT'S NAME:

APPLICANT'S MAILING ADDRESS: _____

(Display address exactly as it should appear on registration) _____

APPLICANT'S ENTITY TYPE: (**Check one** and supply requested information)

Individual - Citizen of (Country):

Partnership - State where organized (Country, if appropriate): _____
Names and Citizenship (Country) of General Partners: _____

Corporation - State (Country, if appropriate) of Incorporation: _____

Other (Specific Nature of Entity and Domicile): _____

GOODS AND/OR SERVICES:

Applicant requests registration of the collective mark shown in the accompanying drawing in the United States Patent and Trademark Office on the Principal Register established by the Act of July 5, 1946 (15 U.S.C. 1051 et. seq., as amended) for the following goods/services (**SPECIFIC GOODS AND/OR SERVICES MUST BE INSERTED HERE**): _____

BASIS FOR APPLICATION: (Check boxes which apply, **but never both the first AND second boxes,** and supply requested information related to each box checked.)

[ ] Applicant is exercising legitimate control over the use of the mark in commerce by its members on or in connection with the above identified goods/services. (15 U.S.C. 1051 (a) and 1054, as amended.) Three specimens showing the mark as used by the members in commerce are submitted with this application.

- Date of first use of the mark in commerce which the U.S. Congress may regulate (for example, interstate or between the U.S. and a foreign country): _____
- Specify the type of commerce: _____
  (for example, interstate or between the U.S. and a specified foreign country)
- Date of first use anywhere (the same as or before use in commerce date):
- Specify manner or mode of use of mark on or in connection with the goods/services: _____
  (for example, trademark is applied to labels, service mark is used in advertisements)

[ ] Applicant has a bona fide intention to exercise legitimate control over the use of the mark in commerce by its members on or in connection with the above-identified good/services. (15 U.S. C. 1051(b) and 1054, as amended.)
- Specify intended manner or mode of use of mark on or in connection with the goods/services: _____
  (for example, trademark will be applied to labels, service mark will be used in advertisements)

[ ] Applicant has a bona fide intention to exercise legitimate control over the use of the mark in commerce by its members on or in connection with the above identified goods/services, and asserts a claim of priority based upon a foreign application in accordance with 15 U.S.C. 1126(d), as amended.
- Country of foreign filing: _____ • Date of foreign filing: _____

[ ] Applicant has a bona fide intention to exercise legitimate control over the use of the mark in commerce by its members on or in connection with the above identified goods/services and, accompanying this application, submits a certification or certified copy of a foreign registration in accordance with 15 U.S.C 1126(e), as amended.
- Country of registration: _____ • Registration number: _____

NOTE: Declaration, on Reverse Side, MUST be Signed

Applicant controls (or, if the application is being filed under 15 U.S.C. 1051(b), applicant intends to control) the use of the mark by the members in th following manner: _____

_____

_____

_____

NOTE: If applicant's bylaws or other written provisions specify the manner of control, or intended manner of control, it will be sufficient to state such bylaws or other written provisions.

DECLARATION

The undersigned being hereby warned that willful false statements and the like so made are punishable by fine or imprisonment, or both, under 18 U.S.C. 1001, and that such willful false statements may jeopardize the validity of the application or any resulting registration, declares that he/she is properly authorized to execute this application on behalf of the applicant; he/she believes the applicant to be the owner of the mark sought to be registered, or, if the application is being filed under 15 U.S.C. 1051(b), he/she believes applicant is entitled to exercise legitimate control over use of the mark in commerce; to the best of his/her knowledge and belief no other person, firm, corporation, or association has the right to use the above identified mark in commerce, either in the identical form thereof or in such near resemblance thereto as to be likely, when used on or in connection with the goods/services of such other person, to cause confusion, or to cause mistake, or to deceive; and that all statements made of his/her own knowledge are true and that all statements made on information and belief are believed to be true.

_____   _____
DATE   SIGNATURE

_____   _____
TELEPHONE NUMBER   PRINT OR TYPE NAME AND POSITION

INSTRUCTIONS AND INFORMATION FOR APPLICANT

TO RECEIVE A FILING DATE, THE APPLICATION <u>MUST</u> BE COMPLETED AND SIGNED BY THE APPLICANT AND SUBMITTED ALONG WITH:

1. The prescribed **FEE ($245.00)** for each class of goods/services listed in the application (**please note that fees are subject to change usually on October 1 of each year)**;
2. A **DRAWING PAGE** displaying the mark in conformance with 37 CFR 2.52;
3. If the application is based on use of the mark in commerce, **THREE (3) SPECIMENS** (evidence) of the mark as used by members in commerce. All three specimens may be the same and may be in the nature of: (a) labels showing the mark which are placed on the goods; (b) photographs of the marks as it appears on the goods, (c) brochures or advertisements showing the mark as used in connection with the services.
4. An **APPLICATION WITH DECLARATION** (this form) - The application must be signed in order for the application to receive a filing date. Only the following persons may sign the declaration: (a) the individual applicant; (b) an officer of the corporate applicant; (c) one general partner of a partnership applicant; (d) all joint applicants.

SEND APPLICATION FORM, DRAWING PAGE, FEE, AND SPECIMENS (IF APPROPRIATE) TO:

Assistant Commissioner for Trademarks
Box New App/Fee
2900 Crystal Drive
Arlington, VA 22202-3513

Additional information concerning the requirements for filing an application is available in a booklet entitled **Basic Facts About Registering a Trademark,** which may be obtained by writing to the above address or by calling: (703) 308-9000.

This form is estimated to take an average of 1 hour to complete, including time required for reading and understanding instructions, gathering necessary information, recordkeeping, and actually providing the information. Any comments on this form, including the amount of time required to complete this form, should be sent to the Office of Management and Organization, U.S. Patent and Trademark Office, U.S. Department of Commerce, Washington, D.C. 20231. Do NOT send completed forms to this address.

COLLECTIVE MEMBERSHIP MARK APPLICATION, PRINCIPAL REGISTER, WITH DECLARATION	MARK (Word(s) and/or Design)	CLASS NO. 200

TO THE ASSISTANT COMMISSIONER FOR TRADEMARKS:

APPLICANT'S NAME:

APPLICANT'S MAILING ADDRESS:

(Display address exactly as it should appear on registration)

APPLICANT'S ENTITY TYPE: (**Check one** and supply requested information)

Individual - Citizen of (Country):

Partnership - State where organized (Country, if appropriate): _____
Names and Citizenship (Country) of General Partners: _____

Corporation - State (Country, if appropriate) of Incorporation:

Other (Specify Nature of Entity and Domicile):

Applicant requests registration of the collective membership mark shown in the accompanying drawing in the United States Patent and Trademark Office on the Principal Register established by the Act of July 5, 1946 (15 U.S.C. 1051 et. seq., as amended) to indicate membership in a(n): _____

(Specify the type or nature of the organization, for example, a social club, labor union, political society, or an association of real estate brokers.)

BASIS FOR APPLICATION: (Check boxes which apply, **but never both the first AND second boxes,** and supply requested information related to each box checked.)

[ ] Applicant is exercising legitimate control over the use of the mark in commerce by its members to indicate membership. (15 U.SC. 1051(a) and 1054 as amended.) Three specimens showing the mark as used by the members in commerce are submitted with this application.
- Date of first use of the mark by members in commerce which the U.S. Congress may regulate (for example, interstate or between the U.S. and a specified foreign country): _____
- Specify the type of commerce: _____
    (for example, interstate or between the U.S. and a specified foreign country)
- Date of first use anywhere (the same as or before use in commerce date): _____
- Specify manner or method of using mark to indicate membership: _____

    (for example, mark is applied to membership cards, certificates, window decals)

[ ] Applicant has a bona fide intention to exercise legitimate control over the use of the mark in commerce by its members to indicate membership. (15 U.S.C. 1051(b) and 1054, as amended.)
- Specify intended manner or method of using mark to indicate membership _____

    (for example, mark is applied to membership cards, certificates, window decals)

[ ] Applicant has a bona fide intention to exercise legitimate control over the use of the mark in commerce by its members to indicate membership, and asserts a claim of priority based upon a foreign application in accordance with 15 U.S.C. 1126(d), as amended.
- Country of foreign filing: _____
- Date of foreign filing: _____

[ ] Applicant has a bona fide intention to exercise legitimate control over the use of the mark in commerce by its members to indicate membership and, accompanying this application, submits a certification or certified copy of a foreign registration in accordance with 15 U.S.C 1126(e), as amended.
- Country of registration: _____
- Registration number: _____

NOTE: Declaration, on Reverse Side, MUST be Signed

PTO Form 4.8
OMB No. 0651-0009 (Exp. 06/30/98) There is no requirement to respond to this collection of information unless a currently valid OMB Number is displayed.

U.S. DEPARTMENT OF COMMERCE/Patent and Trademark Office

Applicant controls (or, if the application is being filed under 15 U.S.C. 1051(b), applicant intends to control) the use of the mark by the members in the following manner: _____

_____

_____

_____

NOTE: If applicant's bylaws or other written provisions specify the manner of control, or intended manner of control, it will be sufficient to state such bylaws or other written provisions.

DECLARATION

The undersigned being hereby warned that willful false statements and the like so made are punishable by fine or imprisonment, or both, under 18 U.S.C. 1001, and that such willful false statements may jeopardize the validity of the application or any resulting registration, declares that he/she is properly authorized to execute this application on behalf of the applicant; he/she believes the applicant to be the owner of the membership mark sought to be registered, or, if the application is being filed under 15 U.S.C. 1051(b), he/she believes applicant is entitled to exercise legitimate control over use of the mark in commerce; to the best of his/her knowledge and belief no other person, firm, corporation, or association has the right to use the above identified mark in commerce, either in the identical form thereof or in such near resemblance thereto as to be likely, when used on or in connection with the goods/services of such other person, to cause confusion, or to cause mistake, or to deceive; and that all statements made of his/her own knowledge are true and that all statements made on information and belief are believed to be true.

_____   _____
DATE   SIGNATURE

_____   _____
TELEPHONE NUMBER   PRINT OR TYPE NAME AND POSITION

INSTRUCTIONS AND INFORMATION FOR APPLICANT

TO RECEIVE A FILING DATE, THE APPLICATION MUST BE COMPLETED AND SIGNED BY THE APPLICANT AND SUBMITTED ALONG WITH:

1. The prescribed **FEE ($245.00)** for each class of goods/services listed in the application (**please note that fees are subject to change usually on October 1 of each year)**;
2. A **DRAWING PAGE** displaying the mark in conformance with 37 CFR 2.52;
3. If the application is based on use of the mark in commerce, **THREE (3) SPECIMENS** (evidence) of the mark as used in commerce for each class of goods/services listed in the application. All three specimens may be in the nature of: (a) labels showing the mark which are placed on the goods; (b) photographs of the mark as it appears on the goods, (c) brochures or advertisements showing the mark as used in connection with the services.
4. An **APPLICATION WITH DECLARATION** (this form) - The application must be signed in order for the application to receive a filing date. Only the following persons may sign the declaration: (a) the individual applicant; (b) an officer of the corporate applicant; (c) one general partner of a partnership applicant; (d) all joint applicants.

SEND APPLICATION FORM, DRAWING PAGE, FEE, SPECIMENS (IF APPROPRIATE) AND COPY OF STANDARDS TO:

Assistant Commissioner for Trademarks
Box New App/Fee
2900 Crystal Drive
Arlington, VA 22202-3513

Additional information concerning the requirements for filing an application is available in a booklet entitled **Basic Facts About Registering a Trademark,** which may be obtained by writing to the above address or by calling: (703) 308-9000.

This form is estimated to take an average of 1 hour to complete, including time required for reading and understanding instructions, gathering necessary information, recordkeeping, and actually providing the information. Any comments on this form, including the amount of time required to complete this form, should be sent to the Office of Management and Organization, U.S. Patent and Trademark Office, U.S. Department of Commerce, Washington, D.C. 20231. Do NOT send completed forms to this address.

CERTIFICATION MARK APPLICATION, PRINCIPAL REGISTER, WITH DECLARATION	MARK (Word(s) and/or Design)
	CLASS [ ] A. Goods [ ] B. Services

TO THE ASSISTANT COMMISSIONER FOR TRADEMARKS:

APPLICANT'S NAME:

APPLICANT'S MAILING ADDRESS: _____

Display address exactly as it should appear on registration) _____

APPLICANT'S ENTITY TYPE: (**Check one** and supply requested information)

	Individual - Citizen of (Country):
	Partnership - State where organized (Country, if appropriate): _____ Names and Citizenship (Country) of General Partners: _____
	Corporation - State (Country, if appropriate) of Incorporation:
	Other (Specify Nature of Entity and Domicile):

GOODS AND/OR SERVICES:

Applicant requests registration of the certification mark shown in the accompanying drawing in the United States Patent and Trademark Office on the Principal Register established by the Act of July 5, 1946 (15 U.S.C. 1051 et. seq., as amended) for the following goods/services (**SPECIFIC GOODS AND/OR SERVICES MUST BE INSERTED HERE**): _____

The certification mark, as used (or, if filing under 15 U.S.C. 1051(b), intended to be used) by authorized persons, certifies (or, if filing under 15 U.S.C. 1051(b), is intended to certify): _____

(for example, a particular regional origin of the goods, a characteristic of the goods or services, that labor was performed by a particular group)

BASIS FOR APPLICATION: (Check boxes which apply, **but never both the first AND second boxes,** and supply requested information related to each box checked.)

[ ]	Applicant is exercising legitimate control over the use of the certification mark in commerce on or in connection with the above identified goods/services. (15 U.S.C. 1051(a) and 1054 as amended.) Three specimens showing the mark as used by authorized persons in commerce are submitted with this application.
	 • Date of first use anywhere of the mark by authorized person in commerce which the U.S. Congress may regulate (for example, interstate or between the U.S. and a specified foreign country): _____
	 • Specify the type of commerce: _____
	     (for example, interstate or between the U.S. and a specified foreign country)
	 • Date of first use anywhere by an authorized person (the same as or before use in commerce date): _____
	 • Specify manner of using mark on or in connection with the goods/services: _____
	     (for example, trademark is applied to labels, service mark is used in advertisements)

[ ]	Applicant has a bona fide intention to exercise legitimate control over the use of the certification mark in commerce on or in connection with the above identified goods/services. (15 U.S.C. 1051(b) and 1054, as amended.) • Specify intended manner or method of using mark on or in connection with the goods/services: _____ (for example, trademark will be applied to labels, service mark will be used in advertisements)
[ ]	Applicant has a bona fide intention to exercise legitimate control over the use of the certification mark in commerce on or in connection with the above identified goods/services, and asserts a claim of priority based upon a foreign application in accordance with 15 U.S.C. 1126(d), as amended. • Country of foreign filing: _____ • Date of foreign filing: _____
[ ]	Applicant has a bona fide intention to exercise legitimate control over the use of the certification mark in commerce on or in connection with the above identified goods/services and, accompanying this application, submits a certification or certified copy of a foreign registration in accordance with 15 U.S.C 1126(e), as amended. • Country of registration: _____ • Registration number: _____

Applicant is not engaged (or, if filing under 15 U.S.C 1051(b), will not engage) in the production or marketing of the goods orservices to which the mark is applied.

The applicant must also provide a copy of the standards the applicant uses to determine whether goods or services will be certified. If the applicant files based on prior use in commerce, this should be provided with this application. In an application filed based on an intent to use in commerce, this should be provided with the Allegation of Use (Amendment to Allege Use/Statement of Use).

DECLARATION

The undersigned being hereby warned that willful false statements and the like so made are punishable by fine or imprisonment, or both, under 18 U.S.C. 1001, and that such willful false statements may jeopardize the validity of the application or any resulting registration, declares that he/she is properly authorized to execute this application on behalf of the applicant; he/she believes the applicant to be the owner of the mark sought to be registered, or, if the application is being filed under 15 U.S.C. 1051(b), he/she believes applicant is entitled to exercise legitimate control over use of the mark in commerce; to the best of his/her knowledge and belief no other person, firm, corporation, or association has the right to use the above identified mark in commerce, either in the identical form thereof or in such near resemblance thereto as to be likely, when used on or in connection with the goods/services of such other person, to cause confusion, or to cause mistake, or to deceive; and that all statements made of his/her own knowledge are true and that all statements made on information and belief are believed to be true.

_____   _____
DATE   SIGNATURE

_____   _____
TELEPHONE NUMBER   PRINT OR TYPE NAME AND POSITION

INSTRUCTIONS AND INFORMATION FOR APPLICANT

TO RECEIVE A FILING DATE, THE APPLICATION <u>MUST</u> BE COMPLETED AND SIGNED BY THE APPLICANT AND SUBMITTED ALONG WITH:

1. The prescribed **FEE ($245.00)** for each class of goods/services listed in the application (**please note that fees are subject to change usually on October 1 of each year)**;
2. A **DRAWING PAGE** displaying the mark in conformance with 37 CFR 2.52;
3. If the application is based on use of the mark in commerce, **THREE (3) SPECIMENS** (evidence) of the mark as used by members in commerce. All three specimens may be the same and may be in the nature of: (a) labels showing the mark which are placed on the goods; (b) photographs of the marks as it appears on the goods, (c) brochures or advertisements showing the mark as used in connection with the services.
4. An **APPLICATION WITH DECLARATION** (this form) - The application must be signed in order for the application to receive a filing date. Only the following person may sign the declaration: (a) the individual applicant; (b) an officer of the corporate applicant; (c) one general partner of a partnership applicant; (d) all joint applicants.

SEND APPLICATION FORM, DRAWING PAGE, FEE, AND SPECIMENS (IF APPROPRIATE) TO:

**Assistant Commissioner for Trademarks
Box New App/Fee
2900 Crystal Drive
Arlington, VA 22202-3513**

Additional information concerning the requirements for filing an application is available in a booklet entitled **Basic Facts About Registering a Trademark,** which may be obtained by writing to the above address or by calling: (703) 308-9000.

This form is estimated to take an average of 1 hour to complete, including time required for reading and understanding instructions, gathering necessary information, recordkeeping, and actually providing the information. Any comments on this form, including the amountof time required to complete this form, should be sent to the Office of Management and Organization, U.S. Patent and Trademark Office, U.S. Department of Commerce, Washington, D.C. 20231. Do NOT send completed forms to this address.

~TRADEMARK/SERVICE MARK ALLEGATION OF USE (Statement of Use/ Amendment to Allege Use) (15 U.S.C. § 1051(c) or (d))~

NOTE: The following form complies with the provisions of the Trademark Law Treaty Implementation Act (TLTIA).

WHEN TO FILE: Before the USPTO will register a mark that was based upon applicant's bona fide intention to use the mark in commerce, the owner must (1) use the mark in commerce; and (2) file an Allegation of Use. The Allegation of Use can only be filed either **on or before** the day the examining attorney approves the mark for publication in the *Official Gazette*; or **on or after** the day the Notice of Allowance is issued. If the Allegation of Use is filed *between* those periods, it will be returned. To avoid return of an untimely Allegation of Use, you can check the status of your application by calling 703-305-8747, or using http://tarr.uspto.gov/

BASIC INSTRUCTIONS

The following form is written in a "scannable" format that will enable the U.S. Patent and Trademark Office (USPTO) to scan paper filings and capture application data automatically using optical character recognition (OCR) technology. Information is to be entered next to identifying data tags, such as <MARK>. OCR software can be programmed to identify these tags, capture the corresponding data, and transmit this data to the appropriate data fields in the Trademark databases, largely bypassing manual data entry processes.

Please enter the requested information in the blank space that appears to the right of each tagged (< >) element. However, do not enter any information immediately after the section headers (the bolded wording appearing in all capital letters). If you need additional space, first, in the space provided on the form, enter "See attached." Then, please use a separate piece of paper on which you first list the data tag (e.g., <Goods and/or Services Not in Use to be **Deleted**>), followed by the relevant information. Some of the information requested *must* be provided. Other information is either required only in certain circumstances, or provided only at your discretion. **Please consult the "Help" section following the form for detailed explanations as to what information should be entered in each blank space.**

To increase the effectiveness of the USPTO scanners, it is recommended that you use a typewriter to complete the form.

For additional information, please see the *Basic Facts about Trademarks* booklet, available at http://www.uspto.gov/web/offices/tac/doc/basic/, or by calling the Trademark Assistance Center at 703-308-9000.

MAILING INFORMATION

Send the completed form; appropriate fee (the filing fee for the Allegation of Use is $100.00 *per class* of goods and/or services, made payable to "The Commissioner of Patent and Trademarks"); and one (1) **SPECIMEN**, showing the mark as currently used in commerce for at least one product or service in each international class covered, to

(Before Approval for Publication):
Fee
AssistantCommissioner for Trademarks
2900 Crystal Drive
Arlington, VA 22202-3513

(After Notice of Allowance):
Box ITU
Fee
Assistant Commissioner for Trademarks
2900 Crystal Drive
Arlington, VA 22202-3513

You may also wish to include a self-addressed stamped postcard with your submission, on which you identify the mark and serial number, and list each item being submitted (e.g., application, fee, specimen, etc.). We will return this postcard to confirm receipt of your submission.

~TRADEMARK/SERVICE MARK ALLEGATION OF USE (Statement of Use/ Amendment to Allege Use) (15 U.S.C. § 1051(c) or (d))~

~To the Assistant Commissioner for Trademarks~

\<TRADEMARK/SERVICEMARK INFORMATION>

\<Mark>

\<Serial Number>

\<APPLICANT INFORMATION>

\<Name>

\<Street>

\<City>

\<State>

\<Country>

\<Zip/Postal Code>

\<NOTICE OF ALLOWANCE INFORMATION>

\<Notice of Allowance>   ~*Enter YES if you are filing the Allegation of Use **after** a Notice of Allowance has issued. If not, enter NO.*~

\<GOODS AND/OR SERVICES INFORMATION>

\<All Goods and/or Services in Application/Notice of Allowance>~*The owner is using the mark in commerce on or in connection with all goods and/or services listed in the application or Notice of Allowance. If not, list in the next section the goods and/or services not in use to be deleted.*~

\<Goods and/or Services Not in Use to be **Deleted**>~*In the following space, list only those goods and/or services (and/or entire classes(es)) appearing in the application or Notice of Allowance for which the owner is **not** using the mark in commerce.* **LEAVE THIS SPACE BLANK IF THE OWNER IS USING THE MARK ON OR IN CONNECTION WITH ALL THE GOODS AND/OR SERVICES LISTED IN THE APPLICATION OR NOTICE OF ALLOWANCE.**~

\<USE INFORMATION>

\<Date of First Use Anywhere>

\<Date of First Use in Commerce>

\<OPTIONAL - REQUEST TO DIVIDE INFORMATION>

\<Request to Divide>   ~*Enter YES if you are you submitting a request to divide with this document. If not, enter NO.*~

<FEE INFORMATION>

$100.00 x <Number of Classes>	= <Total Fees Paid>

<SPECIMEN AND SIGNATURE INFORMATION>

~Applicant requests registration of the above-identified trademark/service mark in the United States Patent and Trademark Office on the Principal Register established by the Act of July 5, 1946 (15 U.S.C. §1051 et seq., as amended). Applicant is the owner of the mark sought to be registered, and is using the mark in commerce on or in connection with the goods/services identified above, as evidenced by the attached specimen(s) showing the mark as currently used in commerce.
(You MUST ATTACH A SPECIMEN showing the mark as currently used in commerce for at least one product or service in each international class covered.)
The undersigned being hereby warned that willful false statements and the like are punishable by fine or imprisonment, or both, under 18 U.S.C. § 1001, and that such willful false statements and the like may jeopardize the validity of this document, declares that he/she is properly authorized to execute this document on behalf of the Owner; and all statements made of his/her own knowledge are true and that all statements made on information and belief are believed to be true.~

~Signature~ _____

<Date Signed>

<Name>

<Title>

<CONTACT INFORMATION>

<Name>

<Company/Firm Name>

<Street>

<City>

<State>

<Country>

<Zip/Postal Code>

<Telephone Number>

<Fax Number>

<e-Mail Address>

<CERTIFICATE OF MAILING> *~Recommended to avoid lateness due to mail delay.~*

~I certify that the foregoing is being deposited with the United States Postal Service as first class mail, postage prepaid, in an envelope addressed to the Assistant Commissioner for Trademarks, 2900 Crystal Drive, Arlington, VA 22202-3513, on~

<Date of Deposit>

~Signature~ _____

<Name>

~REQUEST FOR EXTENSION OF TIME TO FILE A STATEMENT OF USE (15 U.S.C. § 1051(d))~

NOTE: **The following form complies with the provisions of the Trademark Law Treaty Implementation Act (TLTIA).**

WHEN TO FILE: You must file a Statement of Use within six (6) months after the mailing of the Notice of Allowance, UNLESS, within that same period, you submit a request for a six-month extension of time to file the Statement of Use. The request for an extension must:

- be filed in the USPTO within six (6) months after the issue date of the Notice of Allowance, or previously- granted extension period;
- include a verified statement of the applicant's continued bona fide intention to use the mark in commerce;
- specify the goods/services to which the request pertains as they are identified in the Notice of Allowance; and
- include a fee of $100 for each class of goods/services.

You may request five (5) extensions of time. No extensions may extend beyond thirty-six (36) months from the issue date of the Notice of Allowance. The USPTO must receive the second (2nd), third (3rd), fourth (4th), and fifth (5th) extensions within the previously-granted extension period. Do NOT wait until the request for extension has been granted before filing the next request. In addition to the requirements described above, the second (2nd) and subsequent requests must specify applicant's ongoing efforts to use the mark in commerce.

You may submit one (1) extension request during the six-month period in which you file the Statement of Use, unless the granting of this request would extend the period beyond thirty-six (36) months from the issue date of the Notice of Allowance. Instead of specifying ongoing efforts, for this request you should state the belief that the applicant has made valid use of the mark in commerce, as evidenced by the Statement of Use, but that if the USPTO finds the Statement defective, the applicant will require additional time to file a new Statement of Use.

If the original application were based on both Section 1(a) (Use in Commerce) and 1(b) (Intent to Use), this extension request is necessary only for those goods that were based on Section 1(b). You should NOT file an Extension Request (or Statement of Use) for those goods that were filed based on use in commerce.

BASIC INSTRUCTIONS

The following form is written in a "scannable" format that will enable the U.S. Patent and Trademark Office (USPTO) to scan paper filings and capture application data automatically using optical character recognition (OCR) technology. Information is to be entered next to identifying data tags, such as <MARK>. OCR software can be programmed to identify these tags, capture the corresponding data, and transmit this data to the appropriate data fields in the Trademark databases, largely bypassing manual data entry processes.

Please enter the requested information in the blank space that appears to the right of each tagged (< >) element. However, do not enter any information immediately after the section headers (the bolded wording appearing in all capital letters). Some of the information requested *must* be provided. Other information is either required only in certain circumstances, or provided only at your discretion. **Please consult the "Help" section following the form for detailed explanations as to what information should be entered in each blank space.**

To increase effectiveness of the USPTO scanners, it is recommended that you use a typewriter to complete the form.

For additional information, please see the *Basic Facts about Trademarks* booklet, available at http://www.uspto.gov/web/offices/tac/doc/basic/, or by calling the Trademark Assistance Center, at 703-308-9000.

MAILING INFORMATION

Send the completed form; appropriate fee (the filing fee for the Extension Request is $100.00 *per class* of goods and/or services, made payable to "The Commissioner of Patent and Trademarks"); and one (1) **SPECIMEN**, showing the mark as currently used in commerce for at least one product or service in each international class covered, to

> **Box ITU**
> **Fee**
> **Assistant Commissioner for Trademarks**
> **2900 Crystal Drive**
> **Arlington, VA 22202-3513**

You may also wish to include a self-addressed stamped postcard with your submission, on which you identify the mark and serial number, and list each item being submitted (e.g., application, fee, specimen, etc.). We will return this postcard to confirm receipt of your submission.

~REQUEST FOR EXTENSION OF TIME TO FILE A STATEMENT OF USE (15 U.S.C. § 1051(d))~

~To the Assistant Commissioner for Trademarks~

\<TRADEMARK/SERVICEMARK INFORMATION>

\<Mark>

\<Serial Number>

\<APPLICANT INFORMATION>

\<Name>

\<Street>

\<City>

\<State>

\<Country>

\<Zip/Postal Code>

\<NOTICE OF ALLOWANCE INFORMATION>

\<Notice of Allowance Mailing Date>   *~Enter date in the format MM/DD/YYYY.~*

\<GOODS AND/OR SERVICES INFORMATION>

\<All Goods and/or Services in Notice of Allowance>~*The applicant has a continued bona fide intention to use the mark in commerce on or in connection with all the goods and/or services listed in the Notice of Allowance. If not, list in the next section the goods and/or services to be deleted.~*

\<Goods and/or Services to be **Deleted**>~*In following space, list only those goods/services (or entire classes(es)) appearing in the Notice of Allowance for which the applicant does **not** have a continued bona fide intention to use the mark in commerce. **LEAVE THIS SPACE BLANK IF THE APPLICANT DOES HAVE A CONTINUED BONA FIDE INTENTION TO USE THE MARK IN COMMERCE ON OR IN CONNECTION WITH ALL GOODS/ SERVICES LISTED IN THE NOTICE OF ALLOWANCE.~***

\<EXTENSION REQUEST INFORMATION>~

\<Number of Extension Request>   *~Enter which request (1st, 2nd, 3rd, 4th or 5th) this is following the mailing of the Notice of Allowance.~*

\<ONGOING EFFORTS TO USE MARK IN COMMERCE>~*Applies to 2nd, 3rd, 4th & 5th extension requests only.~*

~*The applicant has made the following ongoing efforts to use the mark in commerce on or in connection with those goods and/or services for which use of the mark in commerce has not yet been made.~*

\<Explanation>

PTO Form 1581 (REV 5/99)
OMB Control No. 0651-0009 (Exp. 8/31/2001)

U.S. DEPARTMENT OF COMMERCE/Patent and Trademark Office
There is no requirement to respond to this collection of information
unless a currently valid OMB number is displayed.

\<STATEMENT OF USE SUBMITTED\>~*if applicable*~

\<Additional Time Requested\>~*Enter YES if you believe the applicant has made valid use of the mark in commerce, as evidenced by the Statement of Use submitted with this request. If the Statement of Use does not meet the requirements of 37 C.F.R. 2.88, you request additional time to correct the Statement of Use. If not, enter NO.*~

\<FEE INFORMATION\>

$100.00 x \<Number of Classes\>   = \<Total Filing Fee Paid\>

\<SIGNATURE INFORMATION\>

~*Applicant is entitled to use the mark sought to be registered and has a continued bona fide intention to use the mark in commerce on or in connection with all the goods and/or services listed in the Notice of Allowance. Applicant **requests a six-month extension of time to file the Statement of Use under 37 CFR 2.89.**~*

The undersigned, being hereby warned that willful false statements and the like are punishable by fine or imprisonment, or both, under 18 U.S.C. § 1001, and that such willful false statements and the like may jeopardize the validity of this document, declares that he/she is properly authorized to execute this document on behalf of the Applicant; and all statements made of his/her own knowledge are true and that all statements made on information and belief are believed to be true.~

~Signature~ _____

\<Date Signed\>

\<Name\>

\<Title\>

\<CONTACT INFORMATION\>

\<Name\>

\<Company/Firm Name\>

\<Street\>

\<City\>

\<State\>

\<Country\>

\<Zip/Postal Code\>

\<Telephone Number\>

\<Fax Number\>

\<e-Mail Address\>

\<CERTIFICATE OF MAILING\>~*Recommended to avoid lateness due to mail delay.*~

~I certify that the foregoing is being deposited with the United States Postal Service as first class mail, postage prepaid, in an envelope addressed to the Assistant Commissioner for Trademarks, 2900 Crystal Drive, Arlington, VA 22202-3513, on~

\<Date of Deposit\>

~Signature~ _____

\<Name\>

LINE-BY-LINE HELP INSTRUCTIONS

TRADEMARK/SERVICE MARK INFORMATION

Mark: Enter the word mark in typed form; or, in the case of a design or stylized mark, a brief description of the mark (e.g., "Design of a fanciful cat").
Serial Number: Enter the eight-digit USPTO serial number (e.g., 75/453687).

APPLICANT INFORMATION

Name: Enter the full name of the applicant of the mark, i.e., the name of the individual, corporation, partnership, or other entity that is seeking registration. If a joint venture organized under a particular business name, enter that name. If joint or multiple applicants, enter the name of each of these applicants. If a trust, enter the name of the trustee or trustees. If an estate, enter the name of the executor or executors.
Note: If ownership of the application has changed, you should establish current ownership, either by (1) recording the appropriate document(s) with the USPTO Assignment Branch; or (2) submitting evidence with this declaration, such as a copy of a document transferring ownership from one party to another. To have the USPTO databases reflect the current applicant, you must choose option (1).
Street: Enter the street address or rural delivery route where the applicant is located.
City: Enter the city and/or foreign area designation where the applicant's address is located.
State: Enter the U.S. state or foreign province in which the applicant's address is located.
Country: Enter the country of the applicant's address. If the address is outside the United States, the applicant must appoint a "Domestic Representative" on whom notices or process in proceedings affecting mark may be served.
Zip/Postal Code: Enter the applicant's U.S. Zip code or foreign country postal identification code.

NOTICE OF ALLOWANCE INFORMATION

Notice of Allowance Mailing Date: Enter the date when the USPTO mailed the Notice of Allowance. Please enter the date in the format of MM/DD/YYYY; e.g., 12/03/1999.

GOODS AND/OR SERVICES INFORMATION

All Goods and/or Services in Notice of Allowance: If the applicant does NOT have a continued bona fide intention to use the mark in commerce on or in connection with *all* of the goods and/or services listed in the Notice of Allowance (even if some goods and/or services were originally based on Section 1(a) (Use in Commerce)), complete the next section. Otherwise, we will presume such bona fide intention to use the mark in commerce on or in connection with all of the goods and/or services listed in the Notice of Allowance.
Goods and/or Services to be Deleted: List all goods and/or services (if any) identified in the Notice of Allowance with which the applicant does NOT have a continued bona fide intention to use the mark in commerce; or, specify an entire international class(es), as appropriate (e.g., Classes 9 & 42).

EXTENSION REQUEST INFORMATION

Number of Extension Request: Indicate whether this is the 1st, 2nd, 3rd, 4th or 5th Request for an Extension of Time to File a Statement of Use that you are submitting.

ONGOING EFFORTS TO USE MARK IN COMMERCE

Explanation: For 2nd, 3rd, 4th and 5th Extension Requests only, provide information regarding ongoing efforts the applicant is making to use the mark in commerce with the identified goods and/or services, such as 1) product or service research or development; 2) market research; 3) promotional activities; 4) steps to acquire distributors; and/or
5) steps to obtain required governmental approval, or similar specified activity.

STATEMENT OF USE SUBMITTED

Additional Time Requested: Enter YES if you believe applicant has made valid use of the mark in commerce, as evidenced by the Statement of Use submitted with this request. If the Statement of Use does not meet the requirements of 37 C.F.R. 2.88, you request additional time to correct the Statement of Use. If not, enter NO.

FEE INFORMATION

The filing fee for the Request for Extension of Time to file a Statement of Use is $100.00 *per class*.

Number of Classes: Enter the number of classes (*not* the international class number(s)) to which the Extension Request applies. For example, if the Extension Request applies to Classes 1, 5 and 25, then the number of classes entered would be "3."

Total Filing Fee Paid: Enter fee amount enclosed (either in form of check or money order in U.S. currency, made payable to "Commissioner of Patents and Trademarks"), or to be charged to an already-existing USPTO deposit account.

SIGNATURE INFORMATION

Signature: The appropriate person must sign the form. A person who is properly authorized to sign on behalf of the applicant is: (1) a person with legal authority to bind the applicant; or (2) a person with firsthand knowledge of the facts and actual or implied authority to act on behalf of the applicant; or (3) an attorney who has an actual or implied written or verbal power of attorney from the applicant.

Date Signed: Enter the date the form is signed.

Name: Enter the name of the person signing the form.

Title: Enter the signatory's title, if applicable, e.g., Vice-President, General Partner, etc.

CONTACT INFORMATION

Although this may be the same as provided elsewhere in the document, please enter the following required information for where the USPTO should mail correspondence. (Please note that correspondence will only be mailed to an address in the U.S. or Canada).

Name: Enter the full name of the contact person.

Company/Firm Name: Enter the name of the contact person's company or firm.

Street: Enter the street address or rural delivery route where the contact person is located.

City: Enter the city and/or foreign area designation where the contact person's address is located.

State: Enter the U.S. state or Canadian province in which the contact person's address is located.

Country: Enter the country of the contact person's address.

Zip Code: Enter the U.S. zip code or Canadian postal code.

Telephone Number: Enter the appropriate telephone number.

Fax Number: Enter the appropriate fax number, if available.

e-mail Address: Enter the appropriate e-mail address, if available.

CERTIFICATE OF MAILING

Although optional, use will avoid lateness due to mail delay. The USPTO considers papers timely filed if deposited with the U.S. States Postal Service with sufficient postage as first class mail on or before the due date, *and* accompanied by a signed Certificate of Mailing attesting to timely deposit. The USPTO will look to the date shown on the Certificate of Mailing, rather than date of actual receipt, to determine timeliness.

Date of Deposit: Enter the date of deposit with the United States Postal Service as first class mail.

Signature: The person signing the certificate should have a reasonable basis to expect that the correspondence will be mailed on or before the indicated date.

Name: Enter the name of the person signing the Certificate of Mailing.

Request to Divide

UNITED STATES DEPARTMENT OF COMMERCE
PATENT AND TRADEMARK OFFICE

Applicant

Mark

Trademark Law Office _____

Serial Number

Trademark Attorney _____

Filed _____

Commissioner of Patents and Trademarks
Washington, D.C. 20231

REQUEST TO DIVIDE APPLICATION

The applicant hereby requests that the application identified above be divided as follows:

Please retain in the original application the following goods/services—(use the language of the original application insofar as possible)

Please include in the new, divided application the following goods/services—(They should be different from and should not overlap, those remaining in the original application.)

(check one)

☐ Enclosed is a check in payment of the filing fee for the divided application

☐ The divided application includes all goods or services in a single class presented in the original, parent application; therefore the applicant submits that no filing fee is due or required.

DATED:

Respectfully,

_____

Telephone Number:
Address:

~DECLARATION OF USE OF MARK IN COMMERCE UNDER § 8 (15 U.S.C. § 1058)~

NOTE: The following form complies with the provisions of the Trademark Law Treaty Implementation Act (TLTIA).

WHEN TO FILE: You must file a Section 8 declaration, specimen, and fee on a date that falls on or between the fifth (5th) and sixth (6th) anniversaries of the registration (or, for an extra fee of $100.00 per class, you may file within the six-month grace period following the sixth (6th) anniversary date). You must subsequently file a Section 8 declaration, specimen, and fee on a date that falls on or between the ninth (9th) and tenth (10th) anniversaries of the registration, and each successive ten-year period thereafter (or, for an extra fee of $100.00 per class, you may file within the six-month grace period). *FAILURE TO FILE A SECTION 8 DECLARATION WILL RESULT IN CANCELLATION OF THE REGISTRATION.*
Note: Because the time for filing a ten-year Section 8 declaration coincides with the time for filing a Section 9 renewal request, a combined §§ 8 & 9 form exists. For more information, please see *Basic Facts about Maintaining a Trademark Registration* (for a copy, call the Trademark Assistance Center, at 703-308-9000).

BASIC INSTRUCTIONS

The following form is written in a "scannable" format that will enable the U.S. Patent and Trademark Office (USPTO) to scan paper filings and capture application data automatically using optical character recognition (OCR) technology. Information is to be entered next to identifying data tags, such as <MARK>. OCR software can be programmed to identify these tags, capture the corresponding data, and transmit this data to the appropriate data fields in the Trademark databases, largely bypassing manual data entry processes.

Please enter the requested information in the blank space that appears to the right of each tagged (< >) element. However, do not enter any information immediately after the section headers (the bolded wording appearing in all capital letters). Some of the information requested *must* be provided. Other information is either required only in certain circumstances, or provided only at your discretion. **Please consult the "Help" section following the form for detailed explanations as to what information should be entered in each blank space.**

To increase the effectiveness of the USPTO scanners, it is recommended that you use a typewriter to complete the form.

MAILING INFORMATION

Send the completed form; appropriate fee (The filing fee for the § 8 Declaration is $100.00 per class, made payable to "The Commissioner of Patent and Trademarks"); and any other required materials to:

Box Post Reg
Fee
Assistant Commissioner for Trademarks
2900 Crystal Drive
Arlington, VA 22202-3513

You may also wish to include a self-addressed stamped postcard with your submission, on which you identify the mark and registration number, and list each item being submitted (e.g., declaration, fee, specimen, etc.). We will return this postcard to you to confirm receipt of your submission.

~DECLARATION OF USE OF MARK IN COMMERCE UNDER § 8 (15 U.S.C. § 1058)~

~To the Assistant Commissioner for Trademarks~

<TRADEMARK/SERVICE MARK INFORMATION>

<Mark>

<Registration Number>

<Registration Date>

<OWNER INFORMATION>

<Name>

<Street>

<City>

<State>

<Country>

<Zip/Postal Code>

<DOMESTIC REPRESENTATIVE>~Required ONLY if the owner's address is outside the United States.~

<Name>   ~is hereby appointed the owner's representative

upon whom notice or process in the proceedings affecting the mark may be served.~

<Street>

<City>

<State>

<Zip Code>

<GOODS AND/OR SERVICES INFORMATION>

<All Goods and/or Services in Existing Registration>~The owner is using the mark in commerce on or in connection with all goods and/or services listed in the existing registration. If not, list in the next section the goods and/or services to be deleted.~

<Goods and/or Services Not in Use to be **Deleted**>~In the following space, list only those goods and/or services (or entire classes(es)) appearing in the registration for which the owner is **no longer** using the mark in commerce. **LEAVE THIS SPACE BLANK IF THE OWNER IS USING THE MARK ON OR IN CONNECTION WITH ALL GOODS AND/OR SERVICES LISTED IN THE REGISTRATION.**~

<FEE INFORMATION>

~Section 8 Filing Fee~

$100.00 x <Number of Classes>   = <Filing Fee Due>

~Grace Period Fee: If filing during the six-month grace period, enter § 8 Grace Period Fee~

$100.00 x <Number of Classes>   = <Grace Period Fee Due>

~Filing Fee Due + Grace Period Fee Due~ = <Total Fees Paid>

PTO Form 1583 (REV 5/99)
OMB Control No. 0651-0009 (Exp. 8/31/2001)
188

U.S. DEPARTMENT OF COMMERCE/Patent and Trademark Office
There is no requirement to respond to this collection of information
unless a currently valid OMB number is displayed.

\<SPECIMEN AND SIGNATURE INFORMATION\>

~*The owner is using the mark in commerce on or in connection with the goods/services identified above, as evidenced by the attached specimen(s) showing the mark as currently used in commerce.*
(You MUST ATTACH A SPECIMEN showing the mark as currently used in commerce for at least one product or service in each international class covered.)
The undersigned, being hereby warned that willful false statements and the like are punishable by fine or imprisonment, or both, under 18 U.S.C. § 1001, and that such willful false statements and the like may jeopardize the validity of this document, declares that he/she is properly authorized to execute this document on behalf of the Owner; and all statements made of his/her own knowledge are true and that all statements made on information and belief are believed to be true.~

~Signature~ _____

\<Date Signed\>

\<Name\>

\<Title\>

\<CONTACT INFORMATION\>

\<Name\>

\<Company/Firm Name\>

\<Street\>

\<City\>

\<State\>

\<Country\>

\<Zip/Postal Code\>

\<Telephone Number\>

\<Fax Number\>

\<e-Mail Address\>

\<CERTIFICATE OF MAILING\>~*Recommended to avoid lateness due to mail delay.*~

~I certify that the foregoing is being deposited with the United States Postal Service as first class mail, postage prepaid, in an envelope addressed to the Assistant Commissioner for Trademarks, 2900 Crystal Drive, Arlington, VA 22202-3513, on~

\<Date of Deposit\>

~Signature~ _____

\<Name\>

LINE-BY-LINE HELP INSTRUCTIONS

TRADEMARK/SERVICE MARK INFORMATION

Mark: Enter the word mark in typed form; or, in the case of a design or stylized mark, a brief description of the mark (e.g., "Design of a fanciful cat").

Registration Number: Enter the USPTO registration number.

Registration Date: Enter the date on which the registration was issued.

OWNER INFORMATION

Name: Enter the full name of the **current** owner of the registration, i.e., the name of the individual, corporation, partnership, or other entity that owns the registration. If joint or multiple owners, enter the name of each of these owners.

Note: If ownership of the registration has changed, you must establish current ownership, either by (1) recording the appropriate document(s) with the USPTO Assignment Branch; or (2) submitting evidence with this declaration, such as a copy of a document transferring ownership from one party to another. To have the USPTO databases reflect the current owner, you must choose option (1).

Street: Enter the street address or rural delivery route where the owner is located.

City: Enter the city and/or foreign area designation where the owner's address is located.

State: Enter the U.S. state or foreign province in which the owner's address is located.

Country: Enter the country of the owner's address. If the address is outside the United States, the owner must appoint a "Domestic Representative" on whom notices or process in proceedings affecting the mark may be served. *See* "Domestic Representative" section, below.

Zip/Postal Code: Enter the owner's U.S. Zip code or foreign country postal identification code.

DOMESTIC REPRESENTATIVE

Complete this section **only** if the address of the current owner is outside the U.S. or one of its territories.

Name: Enter the name of the domestic representative.

Street: Enter the street address or rural delivery route where the domestic representative is located.

City: Enter the city where the domestic representative's address is located.

State: Enter the U.S. state in which the domestic representative's address is located.

Zip Code: Enter the U.S. Zip code.

GOODS AND/OR SERVICES INFORMATION

All Goods and/or Services in Existing Registration: If the owner is NOT using the mark in commerce on or in connection with all of the goods/services listed in the registration, complete the next section. Otherwise, we will presume such use on or in connection with ALL goods and/or services.

Goods and/or Services Not In Use to be Deleted: List the goods and/or services (if any) identified in the registration with which the owner is NO LONGER using the mark in commerce; or, specify an entire international class(es), as appropriate (e.g., Classes 9 & 42).

Note: If the owner is not currently using the mark in commerce on or in connection with some or all of the identified goods and/or services, but expects to resume use, *and* the nonuse is due to special circumstances that excuse the nonuse, you must submit a Declaration of Excusable Nonuse under § 8.

FEE INFORMATION

Section 8 Filing Fee: The filing fee for the § 8 Declaration is $100.00 per class.

Number of Classes: Enter the total number of classes (*not* the international class number(s)) to which the § 8 Declaration applies; e.g., if the § 8 Declaration applies to Classes 1, 5 & 25, then enter the number "3."

Filing Fee Due: Enter total of $100.00 multiplied by the number of classes; e.g., $100.00 x 3 = $300.00.

Grace Period Fee: If filed during six-month grace period, a late fee of $100.00 per class must be submitted.

Number of Classes: See above.

Grace Period Fee Due: Enter total of $100.00 multiplied by number of classes; e.g., $100.00 x 3 = $300.00.

Total Fee Paid: Enter the total of the Filing Fee Due plus the Grace Period Fee Due; e.g., $300.00 + $300.00 = $600.00. This amount must either be enclosed (in the form of a check or money order in U.S. currency, made payable to "Commissioner of Patents and Trademarks"), or charged to an already-existing USPTO deposit account.

Note: If the filing is deficient, additional fees may be required.

SPECIMEN AND SIGNATURE INFORMATION

Specimen(s): Attach a specimen showing current use of the registered mark in commerce for at least one product or service in each class that the § 8 Declaration covers; e.g., tags or labels for goods, and advertise-ments for services. Please print the registration number directly on the specimen (or on a label attached thereto). Specimens must be flat, no larger than 8½ inches (21.6 cm.) wide by 11.69 inches (29.7 cm.) long.

Signature: The appropriate person must sign the form. A person who is properly authorized to sign on behalf of the owner is: (1) a person with legal authority to bind the owner; or (2) a person with firsthand knowledge of the facts and actual or implied authority to act on behalf of the owner; or (3) an attorney who has an actual or implied written or verbal power of attorney from the owner.

Date Signed: Enter the date the form is signed.

Name: Enter the name of the person signing the form.

Title: Enter the signatory's title, if applicable, e.g., Vice-President, General Partner, etc.

CONTACT INFORMATION

Although this may be the same as provided elsewhere in the document, please enter the following required information for where the USPTO should mail correspondence. (Please note that correspondence will only be mailed to an address in the U.S. or Canada).

Name: Enter the full name of the contact person.

Company/Firm Name: Enter the name of the contact person's company or firm.

Street: Enter the street address or rural delivery route where the contact person is located.

City: Enter the city and/or foreign area designation where the contact person's address is located.

State: Enter the U.S. state or Canadian province in which the contact person's address is located.

Country: Enter the country of the contact person's address.

Zip Code: Enter the U.S. Zip code or Canadian postal code.

Telephone Number: Enter the appropriate telephone number.

Fax Number: Enter the appropriate fax number, if available.

e-mail Address: Enter the appropriate e-mail address, if available.

CERTIFICATE OF MAILING

Although optional, use of this section is recommended to avoid lateness due to mail delay. Papers are considered timely filed if deposited with the United States Postal Service with sufficient postage as first class mail on or before the due date and accompanied by a signed Certificate of Mailing attesting to timely deposit. The USPTO will look to the date shown on the Certificate of Mailing, rather than the date of actual receipt, to determine the timeliness of this document.

Date of Deposit: Enter the date of deposit with the United States Postal Service as first class mail.

Signature: The person signing the certificate should have a reasonable basis to expect that the correspondence will be mailed on or before the indicated date.

Name: Enter the name of the person signing the Certificate of Mailing.

~DECLARATION OF INCONTESTABILITY OF MARK UNDER §15 (15 U.S.C. § 1065)~

> **NOTE: The following form complies with the provisions of the Trademark Law Treaty Implementation Act (TLTIA).**

> **WHEN TO FILE:** To claim that a mark registered on the Principal Register is now incontestable, you must file a Section 15 declaration once the mark has been in continuous use in commerce for a period of five (5) years after the date of registration, or date of publication under § 12(c), *and* the mark is still in use in commerce. (Section 15 does NOT apply to marks on the Supplemental Register). You may file this declaration within one (1) year after the expiration of either any five-year period of continuous use following registration, or any five-year period of continuous use after publication under § 12(c). For more information, please see *Basic Facts about Maintaining a Trademark Registration* (for a copy, call the Trademark Assistance Center, at 703-308-9000).

BASIC INSTRUCTIONS

> The following form is written in a "scannable" format that will enable the U.S. Patent and Trademark Office (USPTO) to scan paper filings and capture application data automatically using optical character recognition (OCR) technology. Information is to be entered next to identifying data tags, such as <MARK>. OCR software can be programmed to identify these tags, capture the corresponding data, and transmit this data to the appropriate data fields in the Trademark databases, largely bypassing manual data entry processes.
>
> Please enter the requested information in the blank space that appears to the right of each tagged (<   >) element. However, do not enter any information immediately after the section headers (the bolded wording appearing in all capital letters). Some of the information requested *must* be provided. Other information is either required only in certain circumstances, or provided only at your discretion. **Please consult the "Help" section following the form for detailed explanations as to what information should be entered in each blank space.**
>
> To increase the effectiveness of the USPTO scanners, it is recommended that you use a typewriter to complete the form.

MAILING INFORMATION

> Send the completed form; appropriate fee (The filing fee for the § 15 Declaration of Incontestibility is $100.00 per class, made payable to "The Commissioner of Patent and Trademarks"); and any other required materials to:
>
> > Box Post Reg
> > Fee
> > Assistant Commissioner for Trademarks
> > 2900 Crystal Drive
> > Arlington, VA   22202-3513
>
> You may also wish to include a self-addressed stamped postcard with your submission, on which you identify the mark and registration number, and list each item being submitted (e.g., declaration, fee, specimen, etc.). We will return this postcard to you to confirm receipt of your submission.

~DECLARATION OF INCONTESTABILITY OF MARK UNDER §15 (15 U.S.C. § 1065)~

~To the Assistant Commissioner for Trademarks~

<TRADEMARK/SERVICE MARK INFORMATION>

<Mark>

<Registration Number>

<Registration Date>

<OWNER INFORMATION>

<Name>

<Street>

<City>

<State>

<Country>

<Zip/Postal Code>

<GOODS AND/OR SERVICES INFORMATION>

<All Goods and/or Services in Existing Registration>~*The owner has used the mark in commerce for five (5) consecutive years after the date of registration, or the date of publication under § 12(c), and is still using the mark in commerce on or in connection with all goods and/or services listed in the existing registration. If not, list in the next section the goods and/or services not covered.*~

<Goods and/or Services **Not Covered**>~*In the following space, list only those goods and/or services (or entire classes(es)) appearing in the registration for which either the owner has NOT used the mark in commerce for five (5) consecutive years or is NO LONGER using the mark in commerce.* **LEAVE THIS SPACE BLANK IF THE OWNER HAS USED THE MARK IN COMMERCE FOR FIVE (5) CONSECUTIVE YEARS AFTER THE DATE OF REGISTRATION, OR THE DATE OF PUBLICATION UNDER § 12(C), AND IS STILL USING THE MARK IN COMMERCE ON OR IN CONNECTION WITH ALL GOODS/ SERVICES IN THE EXISTING REGISTRATION.**~

U.S. DEPARTMENT OF COMMERCE/Patent and Trademark Office
There is no requirement to respond to this collection of information
unless a currently valid OMB number is displayed.

<FEE INFORMATION>

$100.00 x <Number of Classes>	= <Total Filing Fee Paid>

<SIGNATURE AND OTHER INFORMATION>

~The mark has been in continuous use in commerce for five (5) consecutive years after the date of registration, or the date of publication under § 12(c), and is still in use in commerce on or in connection with all goods and/or services listed in the existing registration. There has been no final decision adverse to the owner's claim of ownership of such mark for such goods and/or services, or to the owner's right to register the same or to keep the same on the register; and there is no proceeding involving said rights pending and not disposed of either in the Patent and Trademark Office or in the courts.~

The undersigned, being hereby warned that willful false statements and the like are punishable by fine or imprisonment, or both, under 18 U.S.C. § 1001, and that such willful false statements and the like may jeopardize the validity of this document, declares that he/she is properly authorized to execute this document on behalf of the Owner; and all statements made of his/her own knowledge are true and that all statements made on information and belief are believed to be true.~

~Signature~_____

<Date Signed>

<Name>

<Title>

<CONTACT INFORMATION>

<Name>

<Company/Firm Name>

<Street>

<City>

<State>

<Country>

<Zip/Postal Code>

<Telephone Number>

<Fax Number>

<e-Mail Address>

<CERTIFICATE OF MAILING>~*Recommended to avoid lateness due to mail delay.*~

~I certify that the foregoing is being deposited with the United States Postal Service as first class mail, postage prepaid, in an envelope addressed to the Assistant Commissioner for Trademarks, 2900 Crystal Drive, Arlington, VA 22202-3513, on~

<Date of Deposit>

~Signature~ _____

<Name>

LINE-BY-LINE HELP INSTRUCTIONS

TRADEMARK/SERVICE MARK INFORMATION

Mark: Enter the word mark in typed form; or, in the case of a design or stylized mark, a brief description of the mark (e.g., "Design of a fanciful cat").

Registration Number: Enter the USPTO registration number.

Registration Date: Enter the date on which the registration was issued.

OWNER INFORMATION

Name: Enter the full name of the current owner of the registration, i.e., the name of the individual, corporation, partnership, or other entity that owns the registration. If joint or multiple owners, enter the name of each of these owners.

Note: If ownership of the registration has changed, you must establish current ownership, either by (1) recording the appropriate document(s) with the USPTO Assignment Branch; or (2) submitting evidence with this declaration, such as a copy of a document transferring ownership from one party to another.

Street: Enter the street address or rural delivery route where the owner is located.

City: Enter the city and/or foreign area designation where the owner's address is located.

State: Enter the U.S. state or foreign province in which the owner's address is located.

Country: Enter the country of the owner's address.

Zip/Postal Code: Enter the owner's U.S. Zip code or foreign country postal identification code.

GOODS AND/OR SERVICES INFORMATION

All Goods and/or Services in Existing Registration: If the owner has NOT used the mark in commerce for five (5) consecutive years after the date of registration, or the date of publication under § 12(c), or is NOT still using the mark in commerce on all the goods and/or services identified in the registration, complete the next section. Otherwise, we will presume such use on or in connection with ALL goods and/or services.

Goods and/or Services Not Covered: List the goods and/or services (if any), or an entire class(es)(e.g., Classes 9 & 42), identified in the registration that the Section 15 does NOT cover, i.e., those goods and/or services, or entire class(es), in connection with which the owner has NOT used the mark in commerce for five (5) consecutive years after the date of registration, or the date of publication under § 12(c), or is no longer using the mark in commerce.

FEE INFORMATION

Section 15 Filing Fee: The fee for the § 15 Declaration of Incontestability is $100.00 per class.

Number of Classes: Enter the total number of classes (*not* the international class number(s)) to which the § 15 Declaration applies. For example, if the § 15 Declaration applies to Classes 1, 5 and 25, then the number "3" should be entered.

Total Filing Fee Paid: Enter the total of the Filing Fee multiplied by the number of classes; e.g., $100.00 x 3 = $300.00. This amount must either be enclosed (in the form of a check or money order in U.S. currency, made payable to "Commissioner of Patents and Trademarks"), or charged to a USPTO deposit account.

REQUIRED SIGNATURE AND OTHER INFORMATION

Signature: The appropriate person must sign the form. A person who is properly authorized to sign on behalf of the owner is: (1) a person with legal authority to bind the owner; or (2) a person with firsthand knowledge of the facts and actual or implied authority to act on behalf of the owner; or (3) an attorney who has an actual or implied written or verbal power of attorney from the owner.

Date Signed: Enter the date the form is signed.

Name: Enter the name of the person signing the form.

Title: Enter the signatory's title, if applicable, e.g., Vice-President, General Partner, etc.

CONTACT INFORMATION

Although this may be the same as provided elsewhere in the document, please enter the following required information for where the USPTO should mail correspondence. (Please note that correspondence will only be mailed to an address in the U.S. or Canada).

Name: Enter the full name of the contact person.

Company/Firm Name: Enter the name of the contact person's company or firm.

Street: Enter the street address or rural delivery route where the contact person is located.

City: Enter the city and/or foreign area designation where the contact person's address is located.

State: Enter the U.S. state or Canadian province in which the contact person's address is located.

Country: Enter the country of the contact person's address.

Zip Code: Enter the U.S. Zip code or Canadian postal code.

Telephone Number: Enter the appropriate telephone number.

Fax Number: Enter the appropriate fax number, if available.

e-mail Address: Enter the appropriate e-mail address, if available.

CERTIFICATE OF MAILING

Although optional, use of this section is recommended to avoid lateness due to mail delay. Papers are considered timely filed if deposited with the United States Postal Service with sufficient postage as first class mail on or before the due date and accompanied by a signed Certificate of Mailing attesting to timely deposit. The USPTO will look to the date shown on the Certificate of Mailing, rather than the date of actual receipt, to determine the timeliness of this document.

Date of Deposit: Enter the date of deposit with the United States Postal Service as first class mail.

Signature: The person signing the certificate should have a reasonable basis to expect that the correspondence will be mailed on or before the indicated date.

Name: Enter the name of the person signing the Certificate of Mailing.

~COMBINED DECLARATION OF USE & INCONTESTIBILITY UNDER §§ 8 & 15 (15 U.S.C. §§ 1058 & 1065)~

NOTE: The following form complies with the provisions of the Trademark Law Treaty Implementation Act (TLTIA).

WHEN TO FILE: You may file a Combined Declaration of Use & Incontestability under Sections 8 & 15 only if you have continuously used a mark registered on the Principal (*not* Supplemental) Register in commerce for five (5) consecutive years after the date of registration. You must file the Combined Declaration, specimen, and fee on a date that falls on or between the fifth and sixth anniversaries of the registration (or, for an extra fee of $100.00 per class, you may file within the six-month grace period following the sixth-anniversary date). If you have NOT continuously used the mark in commerce for five (5) consecutive years, you must *still* file a Section 8 Declaration. You must subsequently file a Section 8 declaration, specimen, and fee on a date that falls on or between the ninth (9th) and tenth (10th) anniversaries of the registration, and each successive ten-year period thereafter (or, for an extra fee of $100.00 per class, you may file within the six-month grace period). *FAILURE TO FILE THE SECTION 8 DECLARATION WILL RESULT IN CANCELLATION OF THE REGISTRATION.* **Note**: Because the time for filing a ten-year Section 8 declaration coincides with the time for filing a Section 9 renewal application, a combined §§ 8 & 9 form exists. For more information, please see *Basic Facts about Maintaining a Trademark Registration* (for a copy, call the Trademark Assistance Center, at 703-308-9000).

BASIC INSTRUCTIONS

The following form is written in a "scannable" format that will enable the U.S. Patent and Trademark Office (USPTO) to scan paper filings and capture application data automatically using optical character recognition (OCR) technology. Information is to be entered next to identifying data tags, such as <MARK>. OCR software can be programmed to identify these tags, capture the corresponding data, and transmit this data to the appropriate data fields in the Trademark databases, largely bypassing manual data entry processes.

Please enter the requested information in the blank space that appears to the right of each tagged (<   >) element. However, do not enter any information immediately after the section headers (the bolded wording appearing in all capital letters). Some of the information requested *must* be provided. Other information is either required only in certain circumstances, or provided only at your discretion. **Please consult the "Help" section following the form for detailed explanations as to what information should be entered in each blank space.**

To increase the effectiveness of the USPTO scanners, it is recommended that you use a typewriter to complete the form.

MAILING INFORMATION

Send the completed form; appropriate fee (The filing fee for Combined Declaration of Use & Incontestibility under §§ 8 & 15 is $200.00 per class, made payable to "The Commissioner of Patent and Trademarks"); and any other required materials to:

> Box Post Reg
> Fee
> Assistant Commissioner for Trademarks
> 2900 Crystal Drive
> Arlington, VA   22202-3513

You may also wish to include a self-addressed stamped postcard with your submission, on which you list each item being submitted (e.g., declaration, fee, specimen, etc.). We will return this postcard to you to confirm receipt of your submission. **199**

~COMBINED DECLARATION OF USE & INCONTESTIBILITY UNDER §§ 8 & 15 (15 U.S.C. §§ 1058 & 1065)~

~To the Assistant Commissioner for Trademarks~

<TRADEMARK/SERVICE MARK INFORMATION>

<Mark

<Registration Number>

<Registration Date>

<OWNER INFORMATION>

<Name>

<Street>

<City>

<State>

<Country>

<Zip/Postal Code>

<DOMESTIC REPRESENTATIVE>~*Required ONLY if the owner's address is outside the United States.*~

<Name>   ~is hereby appointed the owner's representative
upon whom notice or process in the proceedings affecting the mark may be served.~

<Street>

<City>

<State>

<Zip Code>

<GOODS AND/OR SERVICES INFORMATION>

<All Goods and/or Services in Existing Registration> ~*The owner has used the mark in commerce for five (5) consecutive years after the date of registration, or the date of publication under § 12(c), and is still using the mark in commerce on or in connection with all goods and/or services listed in the existing registration. If not, list in the next section the goods and/or services not covered.*~

<Goods and/or Services **Not Covered**>~*In the following space, list only those goods and/or services (or entire classes(es)) appearing in the registration for which either the owner has not used the mark in commerce for five (5) consecutive years or is no longer using the mark in commerce. **LEAVE THIS SPACE BLANK IF THE OWNER HAS USED THE MARK IN COMMERCE FOR FIVE (5) CONSECUTIVE YEARS AFTER THE DATE OF REGISTRATION, OR THE DATE OF PUBLICATION UNDER § 12(C), AND IS STILL USING THE MARK IN COMMERCE ON OR IN CONNECTION WITH ALL THE GOODS AND/OR SERVICES LISTED IN THE EXISTING REGISTRATION.**~*

<FEE INFORMATION>

~Combined §§ 8 & 15 Filing Fee~

$200.00 x <Number of Classes>   = <Filing Fee Due>

~Grace Period Fee: If filing during the six-month grace period, enter § 8 Grace Period Fee~

$100.00 x <Number of Classes>   = <Grace Period Fee Due>

~Filing Fee Due + Grace Period Fee Due = <Total Fees Paid>

<SPECIMEN AND SIGNATURE INFORMATION>

~The owner is using the mark in commerce on or in connection with the goods/services identified above, as evidenced by the attached specimen(s) showing the mark as currently used in commerce. The mark has been in continuous use in commerce for five (5) consecutive years after the date of registration, or the date of publication under § 12(c), and is still in use in commerce on or in connection with all goods and/or services listed in the existing registration. There has been no final decision adverse to the owner's claim of ownership of such mark for such goods and/or services, or to the owner's right to register the same or to keep the same on the register; and there is no proceeding involving said rights pending and not disposed of either in the Patent and Trademark Office or in the courts. **(You MUST ATTACH A SPECIMEN showing the mark as currently used in commerce for at least one product or service in each international class covered.)**

The undersigned, being hereby warned that willful false statements and the like are punishable by fine or imprisonment, or both, under 18 U.S.C. § 1001, and that such willful false statements and the like may jeopardize the validity of this document, declares that he/she is properly authorized to execute this document on behalf of the Owner; and all statements made of his/her own knowledge are true and that all statements made on information and belief are believed to be true.

~Signature~ _____

<Date Signed>

<Name>

<Title>

<CONTACT INFORMATION>

<Name>

<Company/Firm Name>

<Street>

<City>

<State>

<Country>

<Zip/Postal Code>

<Telephone Number>

<Fax Number>

<e-Mail Address>

<CERTIFICATE OF MAILING>*~Recommended to avoid lateness due to mail delay.~*

~I certify that the foregoing is being deposited with the United States Postal Service as first class mail, postage prepaid, in an envelope addressed to the Assistant Commissioner for Trademarks, 2900 Crystal Drive, Arlington, VA 22202-3513, on~

<Date of Deposit>

~Signature~ _____

<Name>

LINE-BY-LINE HELP INSTRUCTIONS

TRADEMARK/SERVICE MARK INFORMATION

Mark: Enter the word mark in typed form; or, in the case of a design or stylized mark, a brief description of the mark (e.g., "Design of a fanciful cat").

Registration Number: Enter the USPTO registration number.

Registration Date: Enter the date on which the registration was issued.

OWNER INFORMATION

Name: Enter the full name of the **current** owner of registration, i.e., the name of the individual, corporation, partnership, or other entity that owns the registration. If joint or multiple owners, enter the name of each of these owners. **Note:** If ownership of the registration has changed, you must establish current ownership, either by (1) recording the appropriate document(s) with the USPTO Assignment Branch; or (2) submitting evidence with this declaration, such as a copy of a document transferring ownership from one party to another. To have the USPTO databases reflect the current owner, you must choose option (1).

Street: Enter the street address or rural delivery route where the owner is located.

City: Enter the city and/or foreign area designation where the owner's address is located.

State: Enter the U.S. state or foreign province in which the owner's address is located.

Country: Enter the country of the owner's address. If the address is outside the United States, the owner must appoint a "Domestic Representative" on whom notices or process in proceedings affecting the mark may be served. *See* "Domestic Representative" section, below.

Zip/Postal Code: Enter the owner's U.S. Zip code or foreign country postal identification code.

DOMESTIC REPRESENTATIVE

Complete this section **only** if the address of the current owner is outside the U.S. or one of its territories.

Name: Enter the name of the domestic representative.

Street: Enter the street address or rural delivery route where the domestic representative is located.

City: Enter the city where the domestic representative's address is located.

State: Enter the U.S. state in which the domestic representative's address is located.

Zip Code: Enter the U.S. Zip code.

GOODS AND/OR SERVICES INFORMATION

All Goods and/or Services in Existing Registration: If the owner has NOT used the mark in commerce for five (5) consecutive years after the date of registration, or the date of publication under § 12(c), or the owner is NOT still using the mark in commerce on all goods/services identified in the registration, complete the next section. Otherwise, we will presume such use on or in connection with ALL goods and/or services.

Goods and/or Services Not Covered: List the goods and/or services (if any), or an entire class(es)(e.g., Classes 9 & 42), identified in the registration that the Combined Sections 8 & 15 does NOT cover, i.e., those goods and/or services, or entire class(es), in connection with which the owner has NOT used the mark in commerce for five (5) consecutive years after the date of registration, or the date of publication under § 12(c), or is no longer using the mark in commerce. **Note:** If the owner is not currently using the mark in commerce on or in connection with some or all of the identified goods/services, but expects to resume use, *and* the nonuse is due to special circumstances that excuse the nonuse, you must submit a Declaration of Excusable Nonuse under § 8.

FEE INFORMATION

Combined Sections 8 & 15 Filing Fee: Filing fee for Combined §§ 8 & 15 Declaration is $200.00 per class.

Number of Classes: Enter the total number of classes (*not* the international class number(s)) to which the §§ 8 & 15 Declaration applies. For example, if the §§ 8 & 15 Declaration applies to Classes 1, 5 and 25, then enter the number "3."

Filing Fee Due: Enter the total of $200.00 multiplied by the number of classes; e.g., $200.00 x 3 = $600.00.
Grace Period Fee: If filed during six-month grace period, a late fee of $100.00 per class must be submitted.
Number of Classes: See above.
Grace Period Fee Due: Enter total of $100.00 multiplied by number of classes; e.g., $100.00 x 3 = $300.00.
Total Fee Paid: Enter the total of the Filing Fee Due plus the Grace Period Fee Due; e.g., $600.00 + $300.00 = $900.00. This amount must either be enclosed (in the form of a check or money order in U.S. currency, made payable to "Commissioner of Patents and Trademarks"), or charged to an already-existing USPTO deposit account.
Note: If the filing is deficient, additional fees may be required.

SPECIMEN AND SIGNATURE INFORMATION

Specimen(s): Attach a specimen showing current use of the registered mark in commerce for at least one product or service in each class that the § 8 Declaration covers; e.g., tags or labels for goods, and advertisements for services. Please print the registration number directly on the specimen (or on a label attached thereto). Specimens must be flat and no larger than 8½ inches (21.6 cm.) wide by 11.69 inches (29.7 cm.) long.
Signature: The appropriate person must sign the form. A person who is properly authorized to sign on behalf of the owner is: (1) a person with legal authority to bind the owner; or (2) a person with firsthand knowledge of the facts and actual or implied authority to act on behalf of the owner; or (3) an attorney who has an actual or implied written or verbal power of attorney from the owner.
Date Signed: Enter the date the form is signed.
Name: Enter the name of the person signing the form.
Title: Enter the signatory's title, if applicable, e.g., Vice-President, General Partner, etc.

CONTACT INFORMATION

Although this may be the same as provided elsewhere in the document, please enter the following required information for where the USPTO should mail correspondence. (Please note that correspondence will only be mailed to an address in the U.S. or Canada).
Name: Enter the full name of the contact person.
Company/Firm Name: Enter the name of the contact person's company or firm.
Street: Enter the street address or rural delivery route where the contact person is located.
City: Enter the city and/or foreign area designation where the contact person's address is located.
State: Enter the U.S. state or Canadian province in which the contact person's address is located.
Country: Enter the country of the contact person's address.
Zip Code: Enter the U.S. Zip code or Canadian postal code.
Telephone Number: Enter the appropriate telephone number.
Fax Number: Enter the appropriate fax number, if available.
e-mail Address: Enter the appropriate e-mail address, if available.

CERTIFICATE OF MAILING

Although optional, use is recommended to avoid lateness due to mail delay. Papers are considered timely filed if deposited with the U.S. Postal Service with sufficient postage as first class mail on or before the due date and accompanied by a signed Certificate of Mailing attesting to timely deposit. The USPTO will look to the date shown on the Certificate of Mailing, rather than the date of actual receipt, to determine timeliness.
Date of Deposit: Enter the date of deposit with the United States Postal Service as first class mail.
Signature: The person signing the certificate should have a reasonable basis to expect that the correspondence will be mailed on or before the indicated date.
Name: Enter the name of the person signing the Certificate of Mailing.

COMBINED DECLARATION OF USE IN COMMERCE/APPLICATION FOR RENEWAL OF REGISTRATION OF MARK UNDER §§ 8 & 9 (15 U.S.C. §§ 1058 & 1059)~

NOTE: The following form complies with the provisions of the Trademark Law Treaty Implementation Act (TLTIA).

WHEN TO FILE: You must file a Section 8 declaration, specimen, and fee on a date that falls on or between the ninth (9th) and tenth (10th) anniversaries of the registration, and each successive ten-year period thereafter (or, for an extra fee of $100.00 per class, you may file within the six-month grace period). Also, you must file a renewal application within the same period (or, for an extra fee of $100.00 per class, you may file within the six-month grace period following the registration expiration date). *FAILURE TO FILE THIS DOCUMENT WILL RESULT IN CANCELLATION/EXPIRATION OF THE REGISTRATION.*
Note: Because the time for filing a ten-year Section 8 declaration coincides with the time for filing a Section 9 renewal application, you may use this combined §§ 8 & 9 form. For more information, please see *Basic Facts about Maintaining a Trademark Registration* (for a copy, call the Trademark Assistance Center, at 703-308-9000).

BASIC INSTRUCTIONS

The following form is written in a "scannable" format that will enable the U.S. Patent and Trademark Office (USPTO) to scan paper filings and capture application data automatically using optical character recognition (OCR) technology. Information is to be entered next to identifying data tags, such as <MARK>. OCR software can be programmed to identify these tags, capture the corresponding data, and transmit this data to the appropriate data fields in the Trademark databases, largely bypassing manual data entry processes.

Please enter the requested information in the blank space that appears to the right of each tagged (< >) element. However, do not enter any information immediately after the section headers (the bolded wording appearing in all capital letters). Some of the information requested *must* be provided. Other information is either required only in certain circumstances, or provided only at your discretion. **Please consult the "Help" section following the form for detailed explanations as to what information should be entered in each blank space.**

To increase the effectiveness of the USPTO scanners, it is recommended that you use a typewriter to complete the form.

MAILING INFORMATION

Send the completed form; appropriate fee (The filing fee for the Combined §§ 8 & 9 Declaration/Application is $400.00, $100.00 per class for the Section 8 Declaration and $300.00 per class for the renewal application, made payable to "The Commissioner of Patent and Trademarks"); and any other required materials to:

> Box Post Reg
> Fee
> Assistant Commissioner for Trademarks
> 2900 Crystal Drive
> Arlington, VA 22202-3513

You may also wish to include a self-addressed stamped postcard with your submission, on which you identify the mark and registration number, and list each item being submitted (e.g., declaration, fee, specimen, etc.). We will return this postcard to confirm receipt of your submission.

COMBINED DECLARATION OF USE IN COMMERCE/APPLICATION FOR RENEWAL OF REGISTRATION OF MARK UNDER §§ 8 & 9 (15 U.S.C. §§ 1058 & 1059)~

~To the Assistant Commissioner for Trademarks~

<TRADEMARK/SERVICE MARK INFORMATION>

<Mark>

<Registration Number>

<Registration Date>

<OWNER INFORMATION>

<Name>

<Street>

<City>

<State>

<Country>

<Zip/Postal Code>

<DOMESTIC REPRESENTATIVE>~Required ONLY if the owner's address is outside the United States.~

<Name>   ~is hereby appointed the owner's
representative upon whom notice or process in the proceedings affecting the mark may be served.~

<Street>

<City>

<State>

<Zip Code>

<GOODS AND/OR SERVICES INFORMATION>

<All Goods and/or Services in Existing Registration>~The owner is using mark in commerce on or in connection with all goods and/or services listed in the existing registration. If not, list in the next section the goods and/or services to be deleted.~

<Goods and/or Services Not in Use to be **Deleted**>~In the following space, list only those goods and/or services (or entire classes(es)) appearing in the registration for which the owner is **no longer** using the mark in commerce. **LEAVE THIS SPACE BLANK IF THE OWNER IS USING THE MARK ON OR IN CONNECTION WITH ALL GOODS AND/OR SERVICES LISTED IN THE REGISTRATION).**~

<FEE INFORMATION>

~Combined §§ 8 & 9 Filing Fee~

$400.00 x <Number of Classes>   = <Filing Fee Due>

~Grace Period Fee: If filing during the six-month grace period, enter Combined §§ 8 & 9 Grace Period Fee.~

$200.00 x <Number of Classes>   = <Grace Fee Due>

~Filing Fee Due + Grace Period Fee Due~ = <Total Fees Paid>

U.S. DEPARTMENT OF COMMERCE/Patent and Trademark Office
There is no requirement to respond to this collection of information
unless a currently valid OMB number is displayed.

<SPECIMEN AND SIGNATURE INFORMATION>

~Section 8: Declaration of Use in Commerce

The owner is using the mark in commerce on or in connection with the goods/services identified above, as evidenced by the attached specimen(s) showing the mark as currently used in commerce.

(You MUST ATTACH A SPECIMEN showing the mark as currently used in commerce for at least one product or service in each international class covered.)

> The undersigned being hereby warned that willful false statements and the like are punishable by fine or imprisonment, or both, under 18 U.S.C. § 1001, and that such willful false statements and the like may jeopardize the validity of this document, declares that he/she is properly authorized to execute this document on behalf of the Owner; and all statements made of his/her own knowledge are true and that all statements made on information and belief are believed to be true.

Section 9: Application for Renewal

The registrant requests that the registration be renewed for the goods and/or services identified above.~

~Signature~ _____

<Date Signed>

<Name>

<Title>

<CONTACT INFORMATION>

<Name>

<Company/Firm Name>

<Street>

<City>

<State>

<Country>

<Zip/Postal Code>

<Telephone Number>

<Fax Number>

<e-Mail Address>

<CERTIFICATE OF MAILING>~*Recommended to avoid lateness due to mail delay.~*

~I certify that the foregoing is being deposited with the United States Postal Service as first class mail, postage prepaid, in an envelope addressed to the Assistant Commissioner for Trademarks, 2900 Crystal Drive, Arlington, VA 22202-3513, on~

<Date of Deposit>

~Signature~ _____

<Name>

DESIGNATION OF DOMESTIC REPRESENTATIVE	MARK *(identify the mark)*
	REGISTRATION NO. (IF KNOWN)
	CLASS NO. (S)

_____

(name of domestic representative)

whose postal address is _____

_____ is hereby designated applicant's representative upon whom

notice or process in proceedings affecting the mark may be served.

_____

(signature of applicant or owner of mark)

_____

(date)

ASSIGNMENT OF **REGISTRATION OF A MARK**	MARK *(identify the mark)*
	REGISTRATION NO. (IF KNOWN)
	CLASS NO. (S)

Whereas _____
(name of assignor)

whose postal address is _____

_____ has adopted, used and is using a mark which is registered in the

United States Patent and Trademark Office, Registration No. _____ dated

_____; and whereas _____
(name of assignee)

whose postal address is _____

is desirous of acquiring said mark and the registration thereof;

Now, therefore, for good and valuable consideration, receipt of which is hereby acknowledged, said

_____ does hereby assign unto the said
(name of assignor)

_____ all right, title and interest in and to
(name of assignee)

the said mark, together with the good will of the business symbolized by the mark, and the above registration

thereof.

_____

*(signature of assignor, if assignor is a corporation or other juristic organization give
the official title of the person who signs for assignor)*

State of _____ } ss.
County of _____

 On this _____ day of _____, 19_____, before me appeared _____

_____ the person who signed this instrument, who acknowledged that he/she

signed it as a free act on his/her own behalf (or on behalf of the identified corporation or other juristic entity

with authority to do so).*

_____

(signature of notary public)

* The wording of the acknowledgment may vary in some jurisdictions. Be sure to use wording acceptable in the jurisdiction where the document is executed.

FORM PTO-1618A
Expires 06/30/99
OMB 0651-0027

U.S. Department of Commerce
Patent and Trademark Office
TRADEMARK

RECORDATION FORM COVER SHEET
TRADEMARKS ONLY

TO: The Commissioner of Patents and Trademarks: Please record the attached original document(s) or copy(ies).

Submission Type

- [ ] **New**
- [ ] **Resubmission** **(Non-Recordation)**
  Document ID # [ ]
- [ ] **Correction of PTO Error**
  Reel # [ ]   Frame # [ ]
- [ ] **Corrective Document**
  Reel # [ ]   Frame # [ ]

Conveyance Type

- [ ] **Assignment**
- [ ] **License**
- [ ] **Security Agreement**
- [ ] **Nunc Pro Tunc Assignment**
- [ ] **Merger**
  Effective Date
  Month   Day   Year
  [ ]
- [ ] **Change of Name**
- [ ] **Other** [ ]

Conveying Party

- [ ] Mark if additional names of conveying parties attached

Execution Date
Month   Day   Year

Name [ ]

Formerly [ ]

- [ ] **Individual**
- [ ] **General Partnership**
- [ ] **Limited Partnership**
- [ ] **Corporation**
- [ ] **Association**
- [ ] **Other** [ ]
- [ ] **Citizenship/State of Incorporation/Organization** [ ]

Receiving Party

- [ ] Mark if additional names of receiving parties attached

Name [ ]

DBA/AKA/TA [ ]

Composed of [ ]

Address (line 1) [ ]

Address (line 2) [ ]

Address (line 3) [ ]
City   State/Country   Zip Code

- [ ] **Individual**
- [ ] **General Partnership**
- [ ] **Limited Partnership**

- [ ] **Corporation**
- [ ] **Association**

If document to be recorded is an assignment and the receiving party is not domiciled in the United States, an appointment of a domestic representative should be attached. *(Designation must be a separate document from Assignment.)*

- [ ] **Other** [ ]
- [ ] **Citizenship/State of Incorporation/Organization** [ ]

FOR OFFICE USE ONLY

Public burden reporting for this collection of information is estimated to average approximately 30 minutes per Cover Sheet to be recorded, including time for reviewing the document and gathering the data needed to complete the Cover Sheet. Send comments regarding this burden estimate to the U.S. Patent and Trademark Office, Chief Information Officer, Washington, D.C. 20231 and to the Office of Information and Regulatory Affairs, Office of Management and Budget, Paperwork Reduction Project (0651-0027), Washington, D.C. 20503. See OMB Information Collection Budget Package 0651-0027, Patent and Trademark Assignment Practice. DO NOT SEND REQUESTS TO RECORD ASSIGNMENT DOCUMENTS TO THIS ADDRESS.

Mail documents to be recorded with required cover sheet(s) information to:
Commissioner of Patents and Trademarks, Box Assignments , Washington, D.C. 20231

FORM PTO-1618B
Expires 06/30/99
OMB 0651-0027

Page 2

U.S. Department of Commerce
Patent and Trademark Office
TRADEMARK

Domestic Representative Name and Address   Enter for the first Receiving Party only.

Name

Address (line 1)

Address (line 2)

Address (line 3)

Address (line 4)

Correspondent Name and Address   Area Code and Telephone Number

Name

Address (line 1)

Address (line 2)

Address (line 3)

Address (line 4)

Pages   Enter the total number of pages of the attached conveyance document including any attachments.

Trademark Application Number(s) or Registration Number(s)   ☐ Mark if additional numbers attached

Enter either the Trademark Application Number or the Registration Number (DO NOT ENTER BOTH numbers for the same property).

Trademark Application Number(s)   Registration Number(s)

Number of Properties   Enter the total number of properties involved.

Fee Amount   Fee Amount for Properties Listed (37 CFR 3.41): $

Method of Payment: Enclosed ☐   Deposit Account ☐
Deposit Account
(Enter for payment by deposit account or if additional fees can be charged to the account.)
Deposit Account Number: #

Authorization to charge additional fees: Yes ☐   No ☐

Statement and Signature

To the best of my knowledge and belief, the foregoing information is true and correct and any attached copy is a true copy of the original document. Charges to deposit account are authorized, as indicated herein.

_____   _____   _____
Name of Person Signing   Signature   Date Signed

GUIDELINES FOR COMPLETING TRADEMARK RECORDATION COVER SHEET

When using this (FORM PTO-1618 A, B, &C), a cover sheet and any necessary continuation sheets must be submitted with each document to be recorded. Enter all required information using standard business block-style print (such as courier 10 pitch). COmpleted cover sheets will be scanned for image capture. Photocopies of the cover sheets are acceptable. Information required for recordation will be extracted from the cover sheet and cover sheet continuation forms only. Submitted cover sheets and documents will become part of the public record. If a document to be recorded concerns both patents and trademarks, a separate patent and a separate trademark cover sheet, including any attached continuing information, must accompany the document. When the document concerns multiple conveyances or transfers, a cover sheet must be submitted for each, if a separate recordation of each transaction is desired. For assistance in completing this cover sheet and information, call 703 308-9723.

Submission Type - Each submission type requires a new cover sheet. Enter an "X" in the appropriate box indicating the type of submission. If the conveyance document is being submitted for recordation for the first time, enter an " X" in the box for New Assignment. If the submission is a Non-recordation, enter an "X" for Re-submission and provide the document identification number of the original submission. Resubmitted non-recordation documents require a <u>new</u> cover sheet (the new cover sheet shall contain all of the appropriate data and the fee required for recordation). If a previously recorded document requires correction due to a data entry error, enter an "X" for Public Correction and provide the reel and frame number of the original document. Requests to correct the data entry error must be submitted on a new cover sheet. The cover sheet shall contain only the data element in question, the name, date and signature of the person submitting the request, and any other pertinent information, (enter the correspondent's name and address, if it has changed since the document was recorded). If a previously recorded document was submitted with erroneous information, enter an "X" indicating Corrective Assignment and provide the reel and frame number of the previously recorded document. A Corrective Assignment requires a <u>new</u> cover sheet as provided in 37CFR 1.334. If the submission type is not listed, enter an "X" in the Other box and specify the submission type.

Conveyance Type - Enter an "X" in the appropriate box describing the nature of the conveying document. If the document is a nunc pro tunc assignment, enter the effective date using the numerical representation of the date without slashes (/) formatted as MMDDYYYY (05141993). If the conveyance type is not listed, enter an "X" in Other Box and specify the nature of the conveyance .

Conveying Party - Enter the full names of all party(ies) conveying the interest. If the conveying party is an individual, enter the last name first, followed by the first name followed by the middle initial. Separate the last and first name by a comma followed by a blank space. For example, "Carter, Constance M." Separate the last and first name by a comma, followed by a blank space. If the conveying party is a corporation and the corporation name begins with "The", the name must be entered as Longmire Cookie Company, The. A Formerly statement. must be entered by placing the word "Formerly " in front of the former business name, separated by a comma (this data is optional). Enter the execution date of the document (i.e. the date the document is signed by each conveying party) using the numerical representation of the month, day, and year without slashes (/) formatted as MMDDYYYY (05141993). Do not use the European date style when entering the date. Indicate the entity and citizenship of each conveying party. If the conveying party is an individual, the country of citizenship must be indicated. If the conveying party is not an individual, then, if it is a U.S. entity, the state under whose laws it is organized should be set out, if it is a foreign entity, the country under whose laws it is organized should be set out. Thus, a U.S. corporation would indicate its state of incorporation, while a foreign corporation would indicate its country of incorporation. The names, execution dates entity and citizenship of additional conveying parties must be entered on the formatted Recordation Form Cover Sheet Continuation. If the entity type is not listed, enter an "X" in the Other Box and specify the entity type. If there are additional conveying parties, enter an "X" in the box indicating additional conveying information is attached. Only the names appearing on the cover sheet and continuation sheets will be recorded.

Receiving Party. - Enter the full name and address of the party(ies) receiving an interest in. If the receiving party is an individual, enter last name first, followed by the first name, followed by the middle initial. Separate the last and first name by a comma, followed by a blank space. If the receiving party is a corporation and the corporation name begins with "The", the name must be entered as" Longmire Cookie Company, The." Indicate the names, and entity of each receiving party. Enter optional information regarding either DBA/AKA/TA , or Composed of, as appropriate. DBA means Doing Business As; AKA means Also Known As; and TA means

Trading As. Enter the appropriate acronym (i.e. DBA, AKA, TA, or Composed of) in front of the business name, separated by a comma. For example, Longmire Cookie Company, The, DBA, Longmire Cookies (this data is optional). Enter up to three lines of address: address line 1 is used to enter the street address; address line 2 is used to enter the floor/room number, suite number or department location; and address line 3 is used to enter the City, State, and zip code. Use the two letter state code when entering the state, (i.e. VA for the state of Virginia). Only the names which appear on this cover sheet and the Recordation Cover Sheet Continuation form(s) will be recorded.

Indicate the entity and citizenship of each receiving party. If the receiving party is an individual, the country of citizenship must be indicated. If the receiving party is not an individual, then, if it is a U.S. entity, the state under whose laws it is organized should be set out, if it is a foreign entity, the country under whose laws it is organized should be set out. Thus a U.S. corporation would indicate its state of incorporation, while a foreign corporation would indicate its country of incorporation (this data is optional). If the document to be recorded is an assignment and the receiving party is not domiciled in the United States, an appointment of domestic representative should be attached. A designation of domestic representative must be contained in a document separate from the assignment document. Enter an "X" in the box to indicate that a designation of domestic representative is attached. If there is more than one party receiving an interest in the property, enter an "X" in the box to indicate that additional information is attached. Only the names which appear on this cover sheet and the Recordation Cover Sheet Continuation form(s) will be recorded.

Correspondent Name and Address - Enter the full name and address of the party to whom correspondence is to be mailed. Each line of address allows up to 40 characters including spaces. Address information will be used to create a mailing label in order to return the document to the submitter. Enter the telephone number and area code of the correspondent.

Number of Pages - Enter the total number of pages contained in the conveyance document, including any attachments. If the document to be recorded concerns both patents and trademarks, separate patent and trademark cover sheets must accompany the document. Do not include the Recordation Form Cover Sheet pages in this total.

Application Numbers or Registration Numbers - Enter the trademark application number(s) (an eight (8) digit number consisting of a two (2) digit series code and a six (6) digit serial number) against which the document is to be recorded. Enter application number(s) as 74105889). **(Do not enter a slash, space or comma between the series code and the serial number).** If an application has matured into a trademark registration, enter the seven digit trademark registration number(s) against which the document is to be recorded. Enter registration numbers as 1714456. **Do not enter both the application number and the registration number for the same property.** Enter application numbers in the space designed for application number(s) and enter registration number(s) in the designated space. Enter property numbers in the designated boxes. Enter an "X" in the appropriate box indicating additional numbers are attached. Enter additional numbers on the Recordation Form Cover Sheet Continuation.

Number of Properties - Enter the total number of applications and registrations identified for recordation including properties indicated on any attached formatted Recordation Form Cover Sheet Continuation(s).

Total Fee Enclosed and Deposit Account Number - A fee is required for each application and patent property against which the document is to be recorded. If the submission concerns multiple conveyances or transfers, a fee must be submitted separately for each property of each conveyance or transfer. Enter the Fee Amount calculated per cover sheet. Enter the Total Fee Enclosed, if payment is made by other than deposit account. If payment is by deposit account, enter the total amount authorized to be charged to the deposit account or merely the "amount due." Enter the deposit account number for authorized charges. Enter an "X" in the Yes or No box indicating authorization to "charge additional fees" to the deposit account. If additional fees are required, the USPTO will generate a request to the USPTO Office of Finance to charge additional fees to the deposit account. A copy of this request will be returned to the submitter with the Notice of Recordation .

Statement and Signature - Enter the name of the person submitting the document. The submitter must sign and date the cover sheet, confirming that to the best of the person's knowledge and belief, the information contained on the cover sheet is correct and that any copy of the document is a true copy of the original document and authorized charges to Deposit Account. The signature and date must appear to the right of the typed name. The document may be signed by the person whose name appears on the documents to be recorded: In the case of an individual, the individual's signature, for a corporation, the signature of an officer, for a partnership, the signature of a general partner, or in any case, the attorney representing such person or entity may sign the document.

RECORDATION FORM COVER SHEET
CONTINUATION
TRADEMARKS ONLY

FORM PTO-1618C
Expires 06/30/99
OMB 0651-0027

U.S. Department of Commerce
Patent and Trademark Office
TRADEMARK

Conveying Party
Enter Additional Conveying Party

☐ Mark if additional names of conveying parties attached

Execution Date
Month Day Year

Name

Formerly

☐ Individual   ☐ General Partnership   ☐ Limited Partnership   ☐ Corporation   ☐ Association

☐ Other

☐ Citizenship State of Incorporation/Organization

Receiving Party
Enter Additional Receiving Party

☐ Mark if additional names of receiving parties attached

Name

DBA/AKA/TA

Composed of

Address (line 1)

Address (line 2)

Address (line 3)

City   State/Country   Zip Code

☐ Individual   ☐ General Partnership   ☐ Limited Partnership

☐ Corporation   ☐ Association

☐ Other

☐ If document to be recorded is an assignment and the receiving party is not domiciled in the United States, an appointment of a domestic representative should be attached *(Designation must be a separate document from the Assignment.)*

☐ Citizenship/State of Incorporation/Organization

Trademark Application Number(s) or Registration Number(s)

☐ Mark if additional numbers attached

Enter either the Trademark Application Number or the Registration Number (DO NOT ENTER BOTH numbers for the same property).

Trademark Application Number(s)   Registration Number(s)

GUIDELINES FOR COMPLETING TRADEMARK RECORDATION COVER SHEET CONTINUATION

Enter any additional information on the Recordation Form Cover Sheet Continuation. Use as many continuation sheets as necessary. Use the same guidelines as appropriate for the Item where the additional data will be entered.

Conveying Party - Enter the full names) of all party(ies) conveying the interest. If the conveying party(ies) is an individual, enter the last name followed by the first name and separated by a comma (i.e. Smith, John). If the conveying party is a corporation and the corporation name begins with "The", the name must be entered as Longmire Cookie Company, The. A Formerly statement. must be entered by placing the word "Formerly " in front of the former business name (this data is optional). Enter the execution date of the document (i.e. the date the document is signed by each conveying party. This date must be entered as the numerical representation of the date without slashes (/) formatted as MMDDYYYY (05141993). Do not use the European date style when entering the date. Indicate the entity and citizenship of each conveying party. If the conveying party is an individual, the country of citizenship must be indicated. If the conveying party is not an individual, then, if it is a U.S. entity, the state under whose laws it is organized should be set out, if it is a foreign entity, the country under whose laws it is organized should be set out. Thus, a U.S. corporation would indicate its state of incorporation, while a foreign corporation would indicate its country of incorporation. The names, execution dates entity and citizenship of additional conveying parties must be entered on the formatted Recordation Form Cover Sheet Continuation. If the entity type is not listed, enter an "X" in the Other Box and specify the entity type. If there are additional conveying parties, enter an "X" in the box indicating additional conveying information is attached. Only the names appearing on the cover sheet and continuation sheets will be recorded.

Receiving Party. - Enter the full name and address of the all parties) receiving an interest in the property. If the receiving party is an individual, enter the last name followed by the first name and separate by a comma (i.e. Smith, John). If the receiving party is a corporation and the corporation name begins with "The", the name must be entered as" Longmire Cookie Company, The." Indicate the names, and entity of each receiving party as well as the execution dates) of the document. Enter optional information regarding either DBA/AKA/TA, or Composed of, as appropriate. DBA means Doing Business As; AKA means Also Known As; and TA means Trading As. Enter the appropriate acronym (i.e. DBA, AKA, TA, or Composed of) in front of the business name. For example, Longmire Cookie Company, The, DBA Longmire Cookies. This data is optional. Enter up to three lines of address: address line 1 is used to enter the floor/room number, suite number or department location; address line 2 is used to enter the street address; and address line 3 is used to enter the City, State, and zip code. Use the two letter state code when entering the state, (i.e. VA for the state of Virginia).

Indicate the entity and citizenship of each receiving party. If the receiving party is an individual, the country of citizenship must be indicated. If the receiving party is not an individual, then, if it is a U.S. entity, the state under whose laws it is organized should be set out, if it is a foreign entity, the country under whose laws it is organized should be set out. Thus a U.S. corporation would indicate its state of incorporation, while a foreign corporation would indicate its country of incorporation. If the document to be recorded is an assignment and the receiving party is not domiciled in the United States, an appointment of domestic representative should be attached. A designation of domestic representative must be contained in a document separate from the assignment document. Enter an "X" in the box to indicate that a designation of domestic representative is attached. If there is more than one party receiving an interest in the property, enter an "X" in the box to indicate that additional information is attached. Only the names appearing on the cover sheet and continuation sheets will be recorded.

Application Numbers or Registration Numbers - Enter the trademark application number (an eight (8) digit number consisting of a two (2) digit series code and a six (6) digit serial number. Enter trademark application numbers as 74105889. **(Do not enter a slash, space or comma between the series code and the serial number).** or trademark registration number (a seven (7) digit number) against which the document is to be recorded. Enter application numbers in the space designed for application number and enter registration numbers in the designated space. If an application has matured into a trademark registration , enter only the registration number. **Do not enter both the application number and the registration number for the same property.** Enter property numbers in the designated boxes (i.e. 1714456 1654123 1682147). Enter an "X" in the appropriate box indicating additional numbers are attached. Enter additional numbers on the Recordation Form Cover Sheet Continuation.

218

Application to Record Trademark
with the United States Customs Service

To: Intellectual Property Rights Branch
U. S. Customs Service
1301 Constitution Ave., N.W.
Washington, DC 20229

Name of trademark owner:

Address of trademark owner:

Trademark owner is:
☐ an individual who is a citizen of _____
☐ a partnership whose partners are citizens of _____
☐ an association or corporation which was organized under the laws of

_____

Places of manufacture of goods bearing the trademark:

The following are foreign persons authorized to use the trademark:

Name:Address: Use authorized:

Identification of any foreign parent or subsidiaries under common ownership or control which uses the trademark abroad*:

Include with this form:
1. A status copy of the certificate of registration certified by the U. S. Patent and Trademark Office showing title to be presently in the name of the applicant.
2. Five copies of the certificate or of a U. S. Patent and Trademark Office Certificate.
3. A fee of $190 for each class of goods sought to be protected.

*Note, "common ownership" means individual or aggregate ownership of more than 50% of the business entity and "common control" means effective control in policy and operations and is not necessarily synonymous with common ownership.

BIBLIOGRAPHY

Hawes, James E. *Trademark Registration Practice*. West Group, 1997.

Kane, Siegrund D. *Trademark Law, A Practitioner's Guide.* New York: Practicing Law Institute, 1997.

Kirkpatrick, Richard L. *Likelihood of Confusion in Trademark Law*. New York: Practicing Law Institute, 1996.

Kramer, Barry and Allen D. Brufsky. *Trademark Law Practice Forms: Rules/Annotations/Commentary*. New York: Clark Boardman Callaghan, 1996.

Levin, William E. *Trade Dress Protection*. West Group.

McCarthy, J. Thomas. McCarthy on Trademark and Unfair Competition. New York: Clark Boardman Callaghan, 1996

Rules of Practice in Trademark Cases. 37 CFR §2.1 et seq.

Vandenburgh, Edward C. *Trademark Law and Practice*. 2d ed. Indianapolis: Bobbs-Merrill Co., Inc., 1968.

U.S. Department of Commerce. *Patent and Trademark Office. Trademark Manual of Examining Procedure*. Second Edition, 1993, Revision 1.1, 1997.

Trademark Act of 1946. 15 USC § 1051 et seq.

Trademark Amendments Act of 1999.

Trademark Law Treaty Implementation Act. Public Law 105-330, October 30, 1998.

INDEX

Your #1 Source for Real World Legal Information...

SPHINX® PUBLISHING
A Division of Sourcebooks, Inc.®

- Written by lawyers
- Simple English explanation of the law
- Forms and instructions included

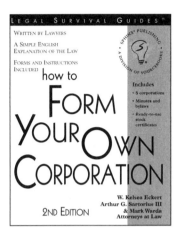

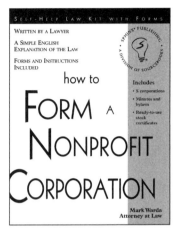

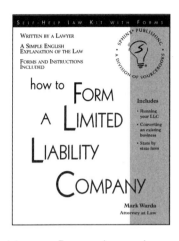

HOW TO FORM YOUR OWN CORPORATION (2ND EDITION)

New business owners can save precious capital by forming their own corporations without the expense of a lawyer. This book includes a summary of the law, forms and instructions for forming a corporation in all 50 states and the District of Columbia.

208 pages; $19.95;
ISBN 1-57071-227-1

HOW TO FORM A NONPROFIT CORPORATION

Forming a nonprofit corporation is not that difficult, and whether you want to start a local school or a national cause group, you can use this book to get the organization started. This book includes all of the forms and instructions needed to form a nonprofit corporation.

192 pages; $24.95;
ISBN 1-57248-099-8

HOW TO FORM A LIMITED LIABILITY COMPANY

Everything you need to start the newest form of doing business. Limited liability companies combine the protection of a corporation with the tax benefits of a partnership. Includes forms and instructions for forming an LLC in all 50 states, with state-by-state law summaries.

192 pages; $19.95;
ISBN 1-57248-083-1

See the following order form for books written specifically for California, Florida, Georgia, Illinois, Massachusetts, Michigan, Minnesota, New York, North Carolina, Pennsylvania, and Texas! *Coming soon—Ohio and New Jersey!*

What our customers say about our books:

"It couldn't be more clear for the lay person." —R.D.

"I want you to know I really appreciate your book. It has saved me a lot of time and money." —L.T.

"Your real estate contracts book has saved me nearly $12,000.00 in closing costs over the past year." —A.B.

"...many of the legal questions that I have had over the years were answered clearly and concisely through your plain English interpretation of the law." —C.E.H.

"If there weren't people out there like you I'd be lost. You have the best books of this type out there." —S.B.

"...your forms and directions are easy to follow." —C.V.M.

Sphinx Publishing's Legal Survival Guides
are directly available from the Sourcebooks, Inc., or from your local bookstores.
For credit card orders call 1–800–43–BRIGHT, write P.O. Box 4410, Naperville, IL 60567-4410,
or fax 630-961-2168

SPHINX® PUBLISHING'S NATIONAL TITLES

Valid in All 50 States

LEGAL SURVIVAL IN BUSINESS

How to Form a Limited Liability Company	$19.95
How to Form Your Own Corporation (2E)	$19.95
How to Form Your Own Partnership	$19.95
How to Register Your Own Copyright (2E)	$19.95
How to Register Your Own Trademark (3E)	$19.95
Most Valuable Business Legal Forms You'll Ever Need (2E)	$19.95
Most Valuable Corporate Forms You'll Ever Need (2E)	$24.95
Software Law (with diskette)	$29.95

LEGAL SURVIVAL IN COURT

Crime Victim's Guide to Justice	$19.95
Debtors' Rights (3E)	$12.95
Defend Yourself against Criminal Charges	$19.95
Grandparents' Rights (2E)	$19.95
Help Your Lawyer Win Your Case (2E)	$12.95
Jurors' Rights (2E)	$9.95
Legal Malpractice and Other Claims against Your Lawyer	$18.95
Legal Research Made Easy (2E)	$14.95
Simple Ways to Protect Yourself from Lawsuits	$24.95
Victims' Rights	$12.95
Winning Your Personal Injury Claim	$19.95

LEGAL SURVIVAL IN REAL ESTATE

How to Buy a Condominium or Townhome	$16.95
How to Negotiate Real Estate Contracts (3E)	$16.95
How to Negotiate Real Estate Leases (3E)	$16.95
Successful Real Estate Brokerage Management	$19.95

LEGAL SURVIVAL IN PERSONAL AFFAIRS

Your Right to Child Custody, Visitation and Support	$19.95
The Nanny and Domestic Help Legal Kit	$19.95
How to File Your Own Bankruptcy (4E)	$19.95
How to File Your Own Divorce (3E)	$19.95
How to Make Your Own Will	$12.95
How to Write Your Own Living Will	$9.95
How to Write Your Own Premarital Agreement (2E)	$19.95
How to Win Your Unemployment Compensation Claim	$19.95
Living Trusts and Simple Ways to Avoid Probate (2E)	$19.95
Neighbor v. Neighbor (2E)	$12.95
The Power of Attorney Handbook (3E)	$19.95
Simple Ways to Protect Yourself from Lawsuits	$24.95
Social Security Benefits Handbook (2E)	$14.95
Unmarried Parents' Rights	$19.95
U.S.A. Immigration Guide (3E)	$19.95
Guia de Inmigracion a Estados Unidos (2E)	$19.95

Legal Survival Guides are directly available from Sourcebooks, Inc., or from your local bookstores.

For credit card orders call 1–800–43–BRIGHT, write P.O. Box 4410, Naperville, IL 60567-4410
or fax 630-961-2168

SPHINX® PUBLISHING ORDER FORM

BILL TO:		SHIP TO:	
Phone #	Terms	F.O.B. Chicago, IL	Ship Date

Charge my: ☐ VISA ☐ MasterCard ☐ American Express

☐ **Money Order or Personal Check**

Credit Card Number

Expiration Date

Qty	ISBN	Title	Retail	Ext.
		SPHINX PUBLISHING NATIONAL TITLES		
	1-57071-166-6	Crime Victim's Guide to Justice	$19.95	
	1-57071-342-1	Debtors' Rights (3E)	$12.95	
	1-57071-162-3	Defend Yourself against Criminal Charges	$19.95	
	1-57248-082-3	Grandparents' Rights (2E)	$19.95	
	1-57248-087-4	Guia de Inmigracion a Estados Unidos (2E)	$19.95	
	1-57248-103-X	Help Your Lawyer Win Your Case (2E)	$12.95	
	1-57071-164-X	How to Buy a Condominium or Townhome	$16.95	
	1-57071-223-9	How to File Your Own Bankruptcy (4E)	$19.95	
	1-57071-224-7	How to File Your Own Divorce (3E)	$19.95	
	1-57248-083-1	How to Form a Limited Liability Company	$19.95	
	1-57248-100-5	How to Form a DE Corporation from Any State	$19.95	
	1-57248-101-3	How to Form a NV Corporation from Any State	$19.95	
	1-57248-099-8	How to Form a Nonprofit Corporation	$24.95	
	1-57071-227-1	How to Form Your Own Corporation (2E)	$19.95	
	1-57071-343-X	How to Form Your Own Partnership	$19.95	
	1-57248-125-0	How to Fire Your First Employee	$19.95	
	1-57248-121-8	How to Hire Your First Employee	$19.95	
	1-57248-119-6	How to Make Your Own Will (2E)	$12.95	
	1-57071-331-6	How to Negotiate Real Estate Contracts (3E)	$16.95	
	1-57071-332-4	How to Negotiate Real Estate Leases (3E)	$16.95	
	1-57248-124-2	How to Register Your Own Copyright (3E)	$19.95	
	1-57248-104-8	How to Register Your Own Trademark (3E)	$19.95	
	1-57071-349-9	How to Win Your Unemployment Compensation Claim	$19.95	
	1-57248-118-8	How to Write Your Own Living Will (2E)	$9.95	
	1-57071-344-8	How to Write Your Own Premarital Agreement (2E)	$19.95	
	1-57071-333-2	Jurors' Rights (2E)	$9.95	
	1-57248-032-7	Legal Malpractice and Other Claims against...	$18.95	
	1-57071-400-2	Legal Research Made Easy (2E)	$14.95	
	1-57071-336-7	Living Trusts and Simple Ways to Avoid Probate (2E)	$19.95	

Qty	ISBN	Title	Retail	Ext.
	1-57071-345-6	Most Valuable Bus. Legal Forms You'll Ever Need (2E)	$19.95	
	1-57071-346-4	Most Valuable Corporate Forms You'll Ever Need (2E)	$24.95	
	1-57248-089-0	Neighbor v. Neighbor (2E)	$12.95	
	1-57071-348-0	The Power of Attorney Handbook (3E)	$19.95	
	1-57248-020-3	Simple Ways to Protect Yourself from Lawsuits	$24.95	
	1-57071-337-5	Social Security Benefits Handbook (2E)	$14.95	
	1-57071-163-1	Software Law (w/diskette)	$29.95	
	0-913825-86-7	Successful Real Estate Brokerage Mgmt.	$19.95	
	1-57248-098-X	The Nanny and Domestic Help Legal Kit	$19.95	
	1-57071-399-5	Unmarried Parents' Rights	$19.95	
	1-57071-354-5	U.S.A. Immigration Guide (3E)	$19.95	
	0-913825-82-4	Victims' Rights	$12.95	
	1-57071-165-8	Winning Your Personal Injury Claim	$19.95	
	1-57248-097-1	Your Right to Child Custody, Visitation and Support	$19.95	
		CALIFORNIA TITLES		
	1-57071-360-X	CA Power of Attorney Handbook	$12.95	
	1-57248-126-9	How to File for Divorce in CA (2E)	$19.95	
	1-57071-356-1	How to Make a CA Will	$12.95	
	1-57071-408-8	How to Probate an Estate in CA	$19.95	
	1-57248-116-1	How to Start a Business in CA	$16.95	
	1-57071-358-8	How to Win in Small Claims Court in CA	$14.95	
	1-57071-359-6	Landlords' Rights and Duties in CA	$19.95	
		FLORIDA TITLES		
	1-57071-363-4	Florida Power of Attorney Handbook (2E)	$12.95	
	1-57248-093-9	How to File for Divorce in FL (6E)	$21.95	
	1-57248-086-6	How to Form a Limited Liability Co. in FL	$19.95	
	1-57071-401-0	How to Form a Partnership in FL	$19.95	
	1-57071-380-4	How to Form a Corporation in FL (4E)	$19.95	
		Form Continued on Following Page	**SUBTOTAL**	

To order, call Sourcebooks at 1-800-43-BRIGHT or FAX (630)961-2168 (Bookstores, libraries, wholesalers—please call for discount)

SPHINX® PUBLISHING ORDER FORM

Qty	ISBN	Title	Retail	Ext.
		FLORIDA TITLES (CONT'D)		
_____	1-57071-361-8	How to Make a FL Will (5E)	$12.95	_____
_____	1-57248-088-2	How to Modify Your FL Divorce Judgment (4E)	$22.95	_____
_____	1-57071-364-2	How to Probate an Estate in FL (3E)	$24.95	_____
_____	1-57248-081-5	How to Start a Business in FL (5E)	$16.95	_____
_____	1-57071-362-6	How to Win in Small Claims Court in FL (6E)	$14.95	_____
_____	1-57071-335-9	Landlords' Rights and Duties in FL (7E)	$19.95	_____
_____	1-57071-334-0	Land Trusts in FL (5E)	$24.95	_____
_____	0-913825-73-5	Women's Legal Rights in FL	$19.95	_____
		GEORGIA TITLES		
_____	1-57071-376-6	How to File for Divorce in GA (3E)	$19.95	_____
_____	1-57248-075-0	How to Make a GA Will (3E)	$12.95	_____
_____	1-57248-076-9	How to Start a Business in Georgia (3E)	$16.95	_____
		ILLINOIS TITLES		
_____	1-57071-405-3	How to File for Divorce in IL (2E)	$19.95	_____
_____	1-57071-415-0	How to Make an IL Will (2E)	$12.95	_____
_____	1-57071-416-9	How to Start a Business in IL (2E)	$16.95	_____
_____	1-57248-078-5	Landlords' Rights & Duties in IL	$19.95	_____
		MASSACHUSETTS TITLES		
_____	1-57071-329-4	How to File for Divorce in MA (2E)	$19.95	_____
_____	1-57248-115-3	How to Form a Corporation in MA	$19.95	_____
_____	1-57248-108-0	How to Make a MA Will (2E)	$12.95	_____
_____	1-57248-109-9	How to Probate an Estate in MA (2E)	$19.95	_____
_____	1-57248-106-4	How to Start a Business in MA (2E)	$16.95	_____
_____	1-57248-107-2	Landlords' Rights and Duties in MA (2E)	$19.95	_____
		MICHIGAN TITLES		
_____	1-57071-409-6	How to File for Divorce in MI (2E)	$19.95	_____
_____	1-57248-077-7	How to Make a MI Will (2E)	$12.95	_____
_____	1-57071-407-X	How to Start a Business in MI (2E)	$16.95	_____
		MINNESOTA TITLES		
_____	1-57248-039-4	How to File for Divorce in MN	$19.95	_____
_____	1-57248-040-8	How to Form a Simple Corporation in MN	$19.95	_____
_____	1-57248-037-8	How to Make a MN Will	$9.95	_____
_____	1-57248-038-6	How to Start a Business in MN	$16.95	_____
		NEW YORK TITLES		
_____	1-57071-184-4	How to File for Divorce in NY	$19.95	_____
_____	1-57248-105-6	How to Form a Corporation in NY	$19.95	_____
_____	1-57248-095-5	How to Make a NY Will (2E)	$12.95	_____
_____	1-57071-185-2	How to Start a Business in NY	$16.95	_____
_____	1-57071-187-9	How to Win in Small Claims Court in NY	$14.95	_____
_____	1-57071-186-0	Landlords' Rights and Duties in NY	$19.95	_____
_____	1-57071-188-7	New York Power of Attorney Handbook	$19.95	_____
_____	1-57248-122-6	Tenants' Rights in NY	$14.95	_____
		NORTH CAROLINA TITLES		
_____	1-57071-326-X	How to File for Divorce in NC (2E)	$19.95	_____
_____	1-57071-327-8	How to Make a NC Will (2E)	$12.95	_____
_____	1-57248-096-3	How to Start a Business in NC (2E)	$16.95	_____
_____	1-57248-091-2	Landlords' Rights & Duties in NC	$19.95	_____
		OHIO TITLES		
_____	1-57248-102-1	How to File for Divorce in OH	$19.95	_____
		PENNSYLVANIA TITLES		
_____	1-57248-127-7	How to File for Divorce in PA (2E)	$19.95	_____
_____	1-57248-094-7	How to Make a PA Will (2E)	$12.95	_____
_____	1-57248-112-9	How to Start a Business in PA (2E)	$16.95	_____
_____	1-57071-179-8	Landlords' Rights and Duties in PA	$19.95	_____
		TEXAS TITLES		
_____	1-57071-330-8	How to File for Divorce in TX (2E)	$19.95	_____
_____	1-57248-009-2	How to Form a Simple Corporation in TX	$19.95	_____
_____	1-57071-417-7	How to Make a TX Will (2E)	$12.95	_____
_____	1-57071-418-5	How to Probate an Estate in TX (2E)	$19.95	_____
_____	1-57071-365-0	How to Start a Business in TX (2E)	$16.95	_____
_____	1-57248-111-0	How to Win in Small Claims Court in TX (2E)	$14.95	_____
_____	1-57248-110-2	Landlords' Rights and Duties in TX (2E)	$19.95	_____

SUBTOTAL THIS PAGE _____

SUBTOTAL PREVIOUS PAGE _____

Illinois residents add 6.75% sales tax

Florida residents add 6% state sales tax plus applicable discretionary surtax

Shipping— $4.00 for 1st book, $1.00 each additional

TOTAL _____